Springer Optimization and Its Applications

VOLUME 51

Managing Editor
Panos M. Pardalos (University of Florida)

Editor–Combinatorial Optimization
Ding-Zhu Du (University of Texas at Dallas)

T0214429

Advisory Board
J. Birge (University of Chicago)
C.A. Floudas (Princeton University)
F. Giannessi (University of Pisa)
H.D. Sherali (Virginia Polytechnic and State University)
T. Terlaky (McMaster University)
Y.Ye (Stanford University)

Aims and Scope
Optimization has been expanding in all directions at an astonishing rate during the last few decades. New algorithmic and theoretical techniques have been developed, the diffusion into other disciplines has proceeded at a rapid pace, and our knowledge of all aspects of the field has grown even more profound. At the same time, one of the most striking trends in optimization is the constantly increasing emphasis on the interdisciplinary nature of the field. Optimization has been a basic tool in all areas of applied mathematics, engineering, medicine, economics, and other sciences.

The series *Springer Optimization and Its Applications* publishes undergraduate and graduate textbooks, monographs, and state-of-the-art expository work that focus on algorithms for solving optimization problems and also study applications involving such problems. Some of the topics covered include nonlinear optimization (convex and nonconvex), network flow problems, stochastic optimization, optimal control, discrete optimization, multiobjective programming, description of software packages, approximation techniques and heuristic approaches.

For further volumes:
http://www.springer.com/series/7393

Mark-Christoph Körner

Minisum Hyperspheres

 Springer

Mark-Christoph Körner
Institut für Numerische und Angewandte Mathematik
Georg-August-Universität Göttingen
Göttingen 37083
Germany
koerner@math.uni-goettingen.de

ISSN 1931-6828
ISBN 978-1-4614-2918-0   ISBN 978-1-4419-9807-1 (eBook)
DOI 10.1007/978-1-4419-9807-1
Springer New York Dordrecht Heidelberg London

Mathematics Subject Classification (2010): 46B20, 52A21, 90B85

Printed on acid-free paper

Springer is part of Springer Science+Business Media (www.springer.com)

Preface

A mathematical problem should be difficult in order to entice us,
yet not completely inaccessible, lest it mock at our efforts.

(David Hilbert)

The most fascinating mathematical problems share two properties. On the one hand, they are easy to understand and can be expressed in simple terms. On the other hand, they are difficult to solve. Probably the best-known example for such a problem is Fermat's Last Theorem: *No three positive integers a, b, and c can satisfy the equation $a^n + b^n = c^n$ for any integer value of n greater than 2.* Whereas Fermat's conjecture is easy to understand, this is certainly not true for the proof given by Andrew Wiles not less than 358 years after Fermat raised his famous conjecture. A less-famed but still fascinating mathematical problem which also goes back to Fermat is the Fermat–Torricelli problem: *Given three fixed points in the plane, find a fourth point such that the sum of distances between the new point and the fixed points is minimal.* A natural generalization of the Fermat–Torricelli problem is the following problem: *Given a set of fixed points in a normed space, find a hypersphere such that the sum of distances between the hypersphere and the fixed points is minimal.* This generalization of the Fermat–Torricelli problem is called *minisum hypersphere problem.* The minisum hypersphere problem can be stated in simple terms but there is no elementary way to solve the problem in general. Hence, the problem fulfills both necessary conditions of a fascinating mathematical problem. Even though these conditions are not sufficient, in my opinion the minisum hypersphere problem is a very interesting mathematical problem. This text presents my research on the minisum hypersphere problem with the intention to inspire the reader for the problem.

A brief outline of the text is given in the following.

- In Chapter 1 a short introduction to the minisum hypersphere problem is given. Related problems and applications of the minisum hypersphere problem are presented. Furthermore, notations are defined and a survey on the mathematical preliminaries is given. Finally, the concept of a *finite dominating set* which provides a bridge between continuous and discrete optimization is introduced.
- Chapter 2 is devoted to the Euclidean version of the minisum hypersphere problem, i.e., hyperspheres are defined with respect to the Euclidean norm and distances between a hypersphere and the fixed points are also measured in the Euclidean norm.

- In Chapter 3 the case is considered where a hypersphere is defined with respect to an arbitrary norm $\| \cdot \|$ and distances are measured in the same norm $\| \cdot \|$. In Section 3.6 the problem under polyhedral norms is discussed for the planar case and a description of optimal solutions is presented.
- In Chapter 4 a slightly different setting is considered. A norm $\| \cdot \|$ in $\mathbb{R}^2$ is used in order to define hyperspheres. But in contrast to Chapters 2 and 3 a different norm k is used to measure distances between a hypersphere and the fixed points. For the class of polyhedral norms, a finite dominating set is identified which gives rise to a solution approach.
- In Chapter 5 the theory developed in Chapter 4 is used in order to analyze the problem of locating a nondegenerated, axis-parallel rectangle. Different variants of this problem are studied. Besides the problem of locating an arbitrary axis-parallel rectangle, problems are discussed where a rectangle has to have a fixed aspect ratio, a fixed perimeter, or both.
- The last chapter of this book, Chapter 6, discusses extensions of the minisum hypersphere problem.

Using notations of set theory, we have the following relations between the chapters of this text:

$$\text{Chapter 2} \subseteq \text{Chapter 3} \qquad\qquad \text{Chapter 3} \cap \text{Chapter 4} \neq \emptyset$$
$$\text{Chapter 4} \cap \text{Chapter 5} \neq \emptyset$$

Chapter 2 is a special case of Chapter 3. In Chapter 4 a generalization of the planar case discussed in Chapter 3 is studied. Furthermore, locating an axis-parallel rectangle with fixed aspect ratio is a special case of the problem analyzed in Chapter 4. (Note that $\subseteq$ and $\cap$ refer to the problems and not to the content.) For instance, not all results of Chapter 2 are also contained in Chapter 3. Therefore, it is recommended to read the text in chronological order.

At this point, I would like to express my gratitude to Prof. Dr. Anita Schöbel for many fruitful discussions, helpful advice and suggestions that helped to improve the presentation of this text. Furthermore, I am indebted to Anita for putting me in touch with Prof. Dr. Jack Brimberg, Prof. Dr. Henrik Juel, and Prof. Dr. Horst Martini. My interest for minisum hyperspheres was sparked by Anita, Jack, and Henrik. I am deeply grateful that they accept me as a member of their research circle. Many parts of this text are based on joint work with them. Special thanks go to Jack and his family for the great time during my research visit in Canada. I would like to thank Prof. Dr. Horst Martini and his assistant Dr. Margarita Spirova for valuable references to related literature. The financial support of the Research Training Group 1023 is also gratefully acknowledged. Furthermore, I thank Dipl.-Kffr. Marina Lehmann and Dr. Michael Schachtebeck for proofreading this text.

Göttingen   Mark-Christoph Körner
September 2010

Contents

Chapter 1
Basic Concepts

1.1 Circles and Hyperspheres

Hyperspheres and circles are mathematical objects which are well known for hundreds of years. The *Rhind Mathematical Papyrus*, written around 1650 BCE by Egyptian mathematicians, already contains a method of approximating the surface area of a circle, see [RS87]. Aristotle (384–322 BCE) argued by observation of stars that the earth must have a spherical shape, see [Dic85]. During course of history, in particular circles were consistently objects of research. For instance, the problem of *squaring the circle* was proposed presumably 500 BCE and remained an open question until von Lindemann proved in 1882 CE the impossibility; see for example [Hob13, Kno93] for a historical survey. But also in the twenty-first century scientific papers still discuss properties of (specific) circles (e.g., [BJS09a, BJS09b]) and hyperspheres (e.g., [BJS07, Nie10]).

In the ancient world, but also nowadays, the terms *circle* and *hypersphere* normally refer to geometric objects which are induced by Euclidean norm

$$\|X\|_2 := \|(x_1,\dots,x_n)\| = \sqrt{|x_1|^2 + \dots + |x_n|^2}.$$

An Euclidean circle $C = C(X,r) \subseteq \mathbb{R}^2$ with center $X \in \mathbb{R}^2$ and radius $r \in \mathbb{R}$ is given by the set

$$C(X,r) = \{Y \in \mathbb{R}^2 : \|X - Y\|_2 = r\}.$$

Analogously, an Euclidean hypersphere $S(X,r)$ in the Euclidean space $\mathbb{R}^n$ is given by the set

$$S(X,r) = \{Y \in \mathbb{R}^n : \|X - Y\|_2 = r\}. \tag{1.1}$$

In this text we extend the concept of circles and hyperspheres to a more general setting. We consider hyperspheres in space $(\mathbb{R}^n, \|\cdot\|)$, where $\|\cdot\|$ is an arbitrary norm in $\mathbb{R}^n$ and $n \geq 2$. Thus, we study hyperspheres in finite dimensional real Banach

M.-C. Körner, *Minisum Hyperspheres*, Springer Optimization and Its Applications 51, DOI 10.1007/978-1-4419-9807-1_1, © Springer Science+Business Media, LLC 2011

spaces, also known as Minkowski spaces [Tho96]. In $(\mathbb{R}^n, \|\cdot\|)$, hyperspheres may be defined analogously to (1.1) by replacing the Euclidean norm $\|\cdot\|_2$ with the norm $\|\cdot\|$.

Definition 1.1. Let $\|\cdot\|$ be a norm in $\mathbb{R}^n$ and $(X,r) \in \mathbb{R}^n \times]0,\infty[$. A hypersphere in $(\mathbb{R}^n, \|\cdot\|)$ is given by the set

$$S(X,r) := \{Y \in \mathbb{R}^n : \|X - Y\| = r\}.$$

X is denoted as *center* and r is denoted as radius of the hypersphere $S(X,r)$. The set of all hyperspheres in $(\mathbb{R}^n, \|\cdot\|)$ is denoted as $\mathscr{G}$. If $\|\cdot\|$ is a norm in the real plane $\mathbb{R}^2$, we refer to the elements of $\mathscr{G}$ by $C(X,r)$ and denote them as *circles*.

Example 1.2. Depending on the space $(\mathbb{R}^n, \|\cdot\|)$, a hypersphere corresponds to different geometric objects. In the Euclidean space $\mathbb{R}^3$ equipped with the Tschebyschow norm $\|\cdot\|_\infty$, spheres correspond to cubes, see Fig. 1.1a. In Fig. 1.1b a diamond is illustrated which corresponds to a circle in the Manhattan plane $(\mathbb{R}^2, \|\cdot\|_1)$.

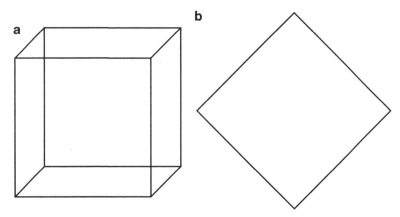

Fig. 1.1 Illustration of Example 1.2. (**a**) Sphere in $(\mathbb{R}^3, \|\cdot\|_\infty)$. (**b**) Circle in $(\mathbb{R}^2, \|\cdot\|_1)$

The distance between a hypersphere and a point is defined in a canonical fashion.

Definition 1.3 (Point–hypersphere distance). The *distance* between a hypersphere $S \subseteq (\mathbb{R}^n, \|\cdot\|)$ and a point $A \in \mathbb{R}^n$ is defined by

$$d(S,A) := \inf\{\|A - Y\| : Y \in S\}.$$

$d(S,A)$ is denoted as *point–hypersphere distance*. Given a set of fixed points $\mathscr{D} = \{A_1, \ldots, A_M\} \subseteq \mathbb{R}^n$, also the shortcut $d_m(S) = d(S,A_m)$, $1 \le m \le M$, is used.

1.2 Minisum Hyperspheres

The problem of finding a minisum hypersphere may be considered as a generalization of the *Weber problem* or the *Fermat–Torricelli problem*. In the following these problems are described briefly. Subsequently, the *minisum hypersphere problem* is defined.

Weber problem.

Given a set of fixed points $\mathscr{D} = \{A_1, \dots, A_M\} \subseteq \mathbb{R}^2$ with positive weights ω_m for each point, find a point $X \in \mathbb{R}^2$ that minimizes the weighted sum of distances between X and the fixed points; that is, find a minimizer of the function

$$f(X) = \sum_{m=1}^{M} \omega_m \|X - A_m\|_2.$$

The denotation *Weber problem* goes back to Alfred Weber (1868–1958). Weber suggested the problem as a model for the optimal location of a facility in the plane (cf. [Web09]). The Weber problem is the corner stone of Location Science and the first location model which has ever been posed in connection with Operations Research (cf. [DKSW02]). Nowadays the Weber problem has been generalized in many ways; for instance higher dimensions, different norms, negative weights, and simultaneous location of multiple facilities have been considered. The Weber problem was also extended to *dimensional facility location* where not a point but a dimensional structure like a straight line, a line segment, or a (convex) set is searched for. The denotation *Weber problem* and the denotations *median problem*, *single facility location problem*, and *minisum problem* are used in Location Science.

Although the Weber problem was named after Alfred Weber, the problem itself is far prior. It is not known who first proposed the Weber problem. But it is behind dispute that Pierre de Fermat (1601–1665) belongs to the head of the list of mathematicians who studied a basic version of the Weber problem. The problem Fermat considered is the following:

Fermat–Torricelli problem.

Given three fixed points in the plane, find a fourth point such that the sum of distances between the new point and the fixed points is minimal.

Evangelista Torricelli (1608–1647) is said to have proposed a first solution to this problem; therefore the problem is named after Fermat and Torricelli. In particular in Geometry also the denotation *Fermat–Torricelli problem* is used in order to refer to the Weber problem. Also the denotation *Steiner–Weber problem* is used, but this denotation is historically wrong (cf. [BSM98]). A more detailed survey on the history of the Weber problem may be found for instance in [DKSW02]. Mathematical properties of the Weber problem in normed spaces are discussed in [MSW02].

In this text we consider the following variant of the Weber problem:

Minisum hypersphere problem.

Let $\mathscr{D} = \{A_1, \ldots, A_M\} \subseteq \mathbb{R}^n$ be a set of fixed points with positive weights $\omega_m > 0$ for all $m = 1, \ldots, M$. Given a norm $\|\cdot\|$ in $\mathbb{R}^n$, find a minimizer of

$$f : \mathscr{G} \to \mathbb{R}, \quad S \mapsto \sum_{m=1}^{M} \omega_m d_m(S)$$

where $d_m(S)$ is the point–hypersphere distance between $A_m \in \mathscr{D}$ and the hypersphere $S \in \mathscr{G}$ (cf. Definition 1.3).

Definition 1.4 (Minisum hypersphere). Let the hypersphere $S \in \mathscr{G}$ be an optimal solution to the minisum hypersphere problem. Then we denote S as *minisum hypersphere*.

Minisum hyperspheres are the object of investigation in this text. Due to the fact that properties of minisum hyperspheres mainly depend on the norm $\|\cdot\|$, we study the minisum hypersphere problem for different types of norms. We start in Chapter 2 with the Euclidean norm. Subsequently, we generalize the Euclidean view and discuss general norms. Finally, we also generalize the minisum hypersphere problem itself and use an arbitrary metric in order to measure the point–hypersphere distance, see Chapter 4. Before we start with our analysis of the minisum hypersphere problem we give a short overview on related problems, discuss applications of the problem, and present some important mathematical preliminaries.

1.2.1 Related Problems

The minisum hypersphere problem itself is mainly studied in Geometry [Nie10] and Location Science [BJS07, BJKS09, KBJS09, KBJS10]. But there are many related problems which have been analyzed in Approximation Theory, Computational Geometry, Measuring Science, and Statistics. In the following we mention a few of them which occur in these fields.

Minimax hypersphere problem.

A problem closely related to the minisum hypersphere problem is the *minimax hypersphere problem*. The minimax hypersphere problem consists of finding a hypersphere S which minimizes the maximum weighted distance from the hypersphere S to the given fixed points. For equal weighted fixed points, any optimal solution to the minimax hypersphere problem induces an annulus which contains all fixed points. In this sense, the minimax hypersphere problem may be interpreted as a covering model. For the Euclidean case this problem has been discussed extensively in the literature. Results for the two-dimensional case may be found in [DSW02, BJS09a]. Higher dimensions are discussed in [GLRS98, Nie02]. In [Cha00] a $(1 + \varepsilon)$-approximation algorithm is presented; [Nie02] contains an exact algorithm which works for any dimension $n \geq 2$. From a geometrical point of view, the most interesting property of the (Euclidean) minimax hypersphere, problem is that for $n = 2$ any optimal solution describes an annulus which has at least two points on the outer circle and at least two points on the inner circle. Also for the minimax hypersphere problem with Manhattan norm results can be found in the literature. In [GHT09] an optimal $\mathcal{O}(M \log M)$ algorithm is presented for the two-dimensional version of the problem. The Euclidean version of the problem is of interest in Measuring Science. There it is used as a model for the *out-of-roundness problem* which occurs in quality control and consists of deciding whether or not the roundness of a part is in the normal range (cf. [FC94, Chapter 14]). Mathematical models for different variants of the out-of-roundness problem are studied for instance in [SLW95, Sun09, LL91, DSW02]. In Location Science the minimax hypersphere problem is also of interest due to the fact that the center X of an optimal hypersphere $S(X, r)$ is a point among all points $Y \in \mathbb{R}^n$ which minimizes the difference

$$\max\{d(Y, A) : A \in \mathscr{D}\} - \min\{d(Y, A) : A \in \mathscr{D}\}$$

where $\mathscr{D} = \{A_1, \ldots, A_M\} \subseteq \mathbb{R}^n$ is a set of fixed points. Therefore the point X may be interpreted as a fair location for a service facility (cf. [Glu08]).

Center problem.

Another problem related to both the minimax and the minisum hypersphere problem consists in finding a smallest hypersphere which contains all fixed points. This problem is equivalent to the well known *center problem* (in Location Science) or *Tschebyschow problem* (in Approximation Theory). An optimal solution to the center or Tschebyschow problem is also denoted as *Tschebyschow center*. The center problem is a point location problem in the sense that given a set $\mathscr{D}$ of fixed points, a new point $X \in \mathbb{R}^n$ is sought after such that the greatest distance between the new point X and the fixed points in $\mathscr{D}$ is minimized. In Location Science in particular the *weighted* two-dimensional version of the problem is well known and a lot of extensions have been considered; an overview is given for instance in [Pla02]. Geometric properties of an optimal solution to the Euclidean two-dimensional center problem are investigated in [DLM07]. They show that the center of a circle is an optimal solution if and only if the circle is rigid in the sense that it cannot be translated while still enclosing all fixed points. The center problem is convex; therefore, general iterative methods of convex programming may be used in order to find an optimal solution to the problem in any dimension $n \geq 2$. For the Euclidean case also a linear time algorithm is known which goes back to Megiddo [Meg82].

Least squares hypersphere problem.

A well-studied variant of the minisum hypersphere problem is the *least squares hypersphere problem* where the distance between the hypersphere S and the fixed points in $\mathcal{D}$ is measured by the squared Euclidean distance. The two-dimensional version of this problem was already studied in 1919 by Coolidge, see [Coo19]. An example for a more recent reference is [CS08] where the least squares hypersphere problem is used as an estimator in a regression model. The existence of optimal solutions to the problem and a perturbation analysis is discussed in [Nie04]. The least squares hypersphere problem is used in particular as a statistical model (see, e.g., [Cha65, BC86, AW04, CLT05, CL05, CS08]). Under the assumption that each fixed point is a noisy observation of some true points which lie on a true Euclidean hypersphere an optimal solution to the least squares hypersphere problem is equivalent to the maximum likelihood estimation of the true hypersphere, provided that all errors in the observation of the true points are normal random variables with zero means and a common variance σ^2 (see, e.g., [Cha65]).

Locating other objects.

Minisum problems are also known for other objects than hyperspheres. The Weber problem mentioned above is for instance a minisum problem where an optimal location of a point is sought. By now also the location of a hyperplane under a minisum objective function is well studied for arbitrary norms (cf. [Sch99, MS01]). For the planar case a review on the location of line segments, half-lines, and polygonal curves under the minisum criterion is given in [DBMS02]. Blanquero et al. suggested in [BCH09] a general solution approach which may be used in order to locate geometric objects like line segments, straight lines, arcs of circumferences, and arbitrary convex sets and their complements or boundaries in the plane $\mathbb{R}^2$. Their approach is based on techniques from DC optimization. In [BJS02] the problem of locating straight lines and line segments in the three-dimensional space $\mathbb{R}^3$ is studied.

1.2.2 Applications

As mentioned before, circles and hyperspheres have been a research topic since the ancient world. No doubt, circles and hyperspheres have aroused interest since time immemorial due to the mathematical beauty that is inherent in them. But the study of properties of circles and hyperspheres was rare ends in themselves. Aristotle for example used his mathematical results in order to justify a thesis on the shape of the earth. Also nowadays applications exist where circles and hyperspheres play a central role. In the previous section we already mentioned a few applications of minimax hyperspheres. In the following we briefly describe an application of minisum hyperspheres in the context of *equity facility location*. Further applications of minisum hyperspheres occur in archeology (analysis of the design and layout of structures), astronomy (identification of the shape of planetary surfaces), computer graphics and vision, electrical engineering (calibration of microwave devices), history (analysis of megalithic monuments), mechanical engineering (measurement of the efficiency of turbines), particle physics (identification of particles in accelerators), and structural engineering (monitoring of deformations); see [Nie10] for an exhaustive list and associated references.

Equity facility location.

Especially in the public sector, efficiency and effectiveness are insufficient criteria in order to determine a location for a new service facility like a sports field, playground, school, and public library, which is accepted by potential users. Instead of that, the location should be fair in the sense that benefit is almost equal for each group within the audience of the facility. But it is not easy to decide whether or not a facility has a fair location. Furthermore, the term *fair* is hard to define. Therefore, in *equity facility location* different measures of equity are discussed, see [MS94] for an overview and [Ogr09, DDG09] for recent works with further references. In order to give an example of a natural measure of equity let $\mathscr{D} \subseteq \mathbb{R}^2$ denote a set of M fixed points which represents the audience of a service facility. Given a possible location $X \in \mathbb{R}^2$ of the new facility and a distance (or utility) measure d the *mean absolute deviation* of X is given by

$$\sum_{A \in \mathscr{D}} \left| d(X,A) - \frac{1}{M} \sum_{A \in \mathscr{D}} d(X,A) \right| . \tag{1.2}$$

The mean absolute deviation measures the sum of deviations from the arithmetic mean of travel distance (or utility). The mean absolute deviation measure is discussed for instance in [BLR76, All78, BK90, Mul91, MS94]. A drawback of this measure is the sensitivity of the arithmetic mean against *outliers*; that is, in the case of equity facility location a small part of the audience of a new service facility may have arbitrarily large impact on a fair location. Thus, the outcome may be considered as unfair by the majority of the audience. An approach to overcome this problem is to replace the mean in (1.2) by the median; that is, a location X will be considered as fair if it minimizes

$$\sum_{A \in \mathscr{D}} |d(X,A) - \text{median}\{d(X,A) : A \in \mathscr{D}\}| . \tag{1.3}$$

In contrast to the arithmetic mean, the median of a set is robust against *outliers* in the sense that moving the locations of up to 50% of the audience arbitrary far away does not affect the value of the median. Therefore, using Equity Measure (1.3) results in a location which is less affected by small groups in the audience with large distance and small benefit, respectively. As mentioned in the previous section, equity location and hypersphere location are related. As it will turn out in Chapter 3, (1.3) is equivalent to a minisum hypersphere problem:

Lemma 1.5. *Let $\mathcal{D} \subseteq \mathbb{R}^n$ be a finite set of fixed points and let $d(X,A) = \|A - X\|$ for all $X,A \in \mathbb{R}^n$.*

- *If $S(X,r) \in \mathcal{G}$ is a minisum hypersphere in $(\mathbb{R}^n, \|\cdot\|)$ with respect to $\mathcal{D}$, then X is a minimizer of (1.3).*
- *If X is a minimizer of (1.3) and $r \in \text{median}\{d(X,A) : A \in \mathcal{D}\}$, then $S(X,r) \in \mathcal{G}$ is a minisum hypersphere with respect to $\mathcal{D}$.*

Thus, among the application listed above, minisum hyperspheres may be applied in order to create fair facility locations. Note that equity and efficiency can be mutually contradictory goals. As an example suppose that all demand points are located on a straight line in the plane $(\mathbb{R}^2, \|\cdot\|_2)$ where $\|\cdot\|_2$ is the Euclidean norm. Then increasing the distance between the fixed points and the new facility may decrease the inequality with respect to measure (1.2) and measure (1.3). In this case a fair location with respect to these measures is given by an idealized point at infinity. Similar observations can be made for minisum hyperspheres as we will see in the following chapters.

1.3 Mathematical Preliminaries

In this section we introduce the notation needed and give a short survey on norms and related concepts. Furthermore we state several results on finite dimensional real Banach spaces which are also known as Minikowski spaces. For a more comprehensive overview on Minkowski spaces we refer to Thompson's textbook on Minkowski Geometry [Tho96] and to the surveys written by Martini et al. [MSW01, MS04].

As in the preceding sections we denote points by capital letters. Furthermore, we use a column vector notation in order to refer to a point $X \in \mathbb{R}^n$. We refer to the transpose of a column vector as X^t. If ambiguity is excluded, then we do not distinguish between column and row vectors. In order to refer to line segments, rays, and straight lines in $\mathbb{R}^n$ we use the following notations.

Notation 1.6. Let $A_1, A_2 \in \mathbb{R}^n$ be two points.

- $[A_1, A_2]$ is the *closed line segment* from A_1 to A_2.
- $]A_1, A_2[$ is the *open line segment* from A_1 to A_2.
- $[A_1, A_1\rangle$ is the *ray* with origin A_1 passing through A_2.
- $\langle A_1, A_2 \rangle$ is the *straight line* through A_1 and A_2.

Several different definitions can be found for the term *hyperplane* in the literature. Within this text *hyperplane* is used in order to refer to an affine subspace of $\mathbb{R}^n$ having dimension $n - 1$. This concept of a hyperplane is equivalent to the following definition.

Definition 1.7. Let $S \in \mathbb{R}^n$ and $b \in \mathbb{R}$. The set

$$H_{S,b} := \{X \in \mathbb{R}^n : S^t X = b\}$$

is called *hyperplane*, provided that $H_{S,b}$ is not empty.

Given a finite set of fixed points and a hypersphere it is convenient to define the sets of points outside, on, and inside the hypersphere.

Notation 1.8. Let $S = S(X, r) \subseteq (\mathbb{R}^n, \|\cdot\|)$ be a hypersphere and $\mathscr{D} \subseteq \mathbb{R}^n$ a finite set of fixed points. We define

$$J_+ = \{A \in \mathscr{D} : \|A - X\| > r\},$$
$$J_0 = \{A \in \mathscr{D} : \|A - X\| = r\},$$
$$J_- = \{A \in \mathscr{D} : \|A - X\| < r\}.$$

Within this text the following definition of *general position* of a set of points in $\mathbb{R}^n$ is used.

Definition 1.9 (General position). Let $U \subseteq \mathbb{R}^n$. Then U is said to be in *general position* if no $n + 1$ of points in U lie in a hyperplane.

We continue with two elementary mathematical concepts.

Definition 1.10. Let $\mathscr{X}$ be a set. A *metric* on $\mathscr{X}$ is a mapping

$$d : \mathscr{X} \times \mathscr{X} \to \mathbb{R}, \quad (X, Y) \mapsto d(X, Y)$$

with the following properties:

  (i) $d(X, Y) = 0$ if and only if $X = Y$.
 (ii) $d(X, Y) = d(Y, X)$ for all $X, Y \in \mathscr{X}$.
(iii) $d(X, Z) \le d(X, Y) + d(Y, Z)$ for all $X, Y, Z \in \mathscr{X}$.

Metrics have some properties which are useful to model distances in the common (nonmathematical) sense: Property (i) is that the distance between two points is zero if and only if the two points are equal. Property (ii) is that the distance between two points is symmetric. The third property of a metric is the well-known *triangle inequality* and means that the direct connection between two points is always the fastest connection. Due to Properties (i)–(iii) it follows that a metric is nonnegative; that is, $d(X, Y) \ge 0$ for all $X, Y \in \mathscr{X}$. A *metric space* is an ordered pair $(\mathscr{X}, d)$ where $\mathscr{X}$ is a set and d is a metric on $\mathscr{X}$. The *diameter* of a subset of a metric space is defined in the following way.

Definition 1.11. Let $(\mathscr{X}, d)$ be a metric space and $\mathscr{A} \subset \mathscr{D}$ a nonempty subset. The *diameter* of $\mathscr{A}$ is defined as

$$\text{diam}(\mathscr{A}) = \sup\{d(X, Y) : X, Y \in \mathscr{A}\}.$$

A norm may be defined in the following way.

Definition 1.12. Let $\mathscr{X}$ be a real vector space. A *norm* $\|\cdot\|$ in X is a mapping

$$\|\cdot\| : \mathscr{X} \to \mathbb{R}, \ X \mapsto \|X\|$$

with the following properties:

(i) $\|X\| = 0$ if and only if $X = 0$.
(ii) $\|\lambda X\| = |\lambda| \|X\|$ for all $\lambda \in \mathbb{R}$ and $X \in \mathscr{X}$.
(iii) $\|X + Y\| \le \|X\| + \|Y\|$ for all $X, Y \in \mathscr{X}$.

Similar to a metric, also property (iii) of a norm is called *triangle inequality*. For collinear points $X, Y \in \mathbb{R}^n$ the triangle inequality holds with equality in any norm, provided that $Y = \lambda X$ for some nonnegative scalar λ.

Lemma 1.13. *Let $\|\cdot\|$ be a norm in $\mathbb{R}^n$, $V \in \mathbb{R}^n$, and denote the origin of $\mathbb{R}^n$ as O. For any $X, Y \in [O, V\rangle$ we have*

$$\|X + Y\| = \|X\| + \|Y\|.$$

Proof. $X, Y \in [O, V\rangle$ imply that there exists $\lambda > 0$ such that $Y = \lambda X$. Therefore, we obtain

$$\|X\| + \|Y\| = \|X\| + \lambda \|X\| = \|(1 + \lambda)X\| = \|X + Y\|.$$

It is well known that each norm induces a metric. $\square$

Lemma 1.14. *Let $(\mathscr{X}, \|\cdot\|)$ be a real normed vector space. Then a metric on $\mathscr{X}$ is defined by*

$$d(X, Y) := \|Y - X\|;$$

d is said to be induced by $\|\cdot\|$.

From Lemma 1.13 it may be concluded that for collinear points $X, Y, Z \in \mathbb{R}^n$ the triangle inequality holds with equality in any metric induced by a norm.

Corollary 1.15. *Let d be a metric which is induced by a norm $\|\cdot\|$ in $\mathbb{R}^n$. Furthermore, let $X, Y, Z \in \mathbb{R}^n$ be collinear and $Z \in [X, Y]$. Then we have*

$$d(X, Y) = d(X, Z) + d(Z, Y).$$

Proof. Let $\lambda \in [0, 1]$ such that $Z = \lambda X + (1 - \lambda)Y$ and define $U := Z - X$, $V := Y - Z$. If $Z = X$ or $Z = Y$ we have nothing to show. Therefore, we assume that $0 < \lambda < 1$. We obtain $U = (1 - \lambda)/\lambda V$. Hence, applying Lemma 1.13 to U and V results in

$$d(X, Y) = \|Y - X\| = \|U + V\| = \|U\| + \|V\|$$
$$= \|Z - X\| + \|Y - Z\| = d(X, Z) + d(Z, Y).$$

$\square$

Note that there does not exist a one-to-one correspondence between norms and metrics. In fact, metrics exist which are not induced by norms, see, e.g., [KS10] for an operations research model where such metrics occur.

There exists an interesting one-to-one correspondence between norms and symmetric compact convex sets: Let $B \subseteq \mathbb{R}^n$ be a full-dimensional compact convex set which is symmetric with respect to the origin and consider the mapping

$$\gamma_B : \mathbb{R}^n \to \mathbb{R}, \quad X \mapsto \gamma_B(X) = \inf\{\lambda > 0 : X \in \lambda B\}. \tag{1.4}$$

Then we have the following result which dates back to Minkowski (see [Min11]).

Lemma 1.16. *The following hold:*

1. Let γ_B be defined as in (1.4). Then γ_B is a norm in $\mathbb{R}^n$.
2. Let $\|\cdot\|$ be a norm in $\mathbb{R}^n$. Then the set

$$B := \{Y \in \mathbb{R}^n : \|Y\| \leq 1\} \tag{1.5}$$

is a full-dimensional compact convex set which is symmetric with respect to the origin.

The set B defined in (1.5) is called *unit ball* of the norm $\|\cdot\|$. Obviously, we have $\|Y\| < 1$ if and only if Y belongs to the interior of B; $\|Y\| = 1$ implies that Y is a point on the boundary of B.

The shape of the unit ball of a norm may be used in order to distinguish between different types of norms. We distinguish between three broad classes of norms in finite dimensional real Banach space: *smooth norms*, *strictly convex norms*, and *polyhedral norms*. These norms are defined as follows.

Definition 1.17. Let $\mathscr{X}$ be a finite dimensional real Banach space and let $\|\cdot\|$ be a norm in $\mathscr{X}$.

- $\|\cdot\|$ is called *smooth norm* if its unit ball has no sharp corners. More formally, $\|\cdot\|$ is a smooth norm if for every $X \in B$ there is a unique supporting hyperplane for B at X, see, e.g., [Phe89, Haz90].
- $\|\cdot\|$ is called a *strictly convex norm* if the boundary of its unit ball contains no line segment, see, e.g., [Thi87, Haz90].
- $\|\cdot\|$ is called *polyhedral norm* or *block norm* if its unit ball is a polytope, see [WWR85, Wit64].

Notice that the class of smooth norms and the class of strictly convex norms are not disjoint. The L_p norm with $1 < p < \infty$, or more specifically the Euclidean norm, is an example of a smooth and strictly convex norm. See Fig. 1.2 and also [NP05, Chapter 2] for an illustrative comparison between smooth, strictly convex, and polyhedral norms.

Remark 1.18. The classification of norms into the classes *smooth*, *strictly convex*, and *polyhedral* is not complete in the sense that norms exist which do not belong to any of theses classes. However, throughout this text we are primarily concerned with norms that belong to at least one of these classes.

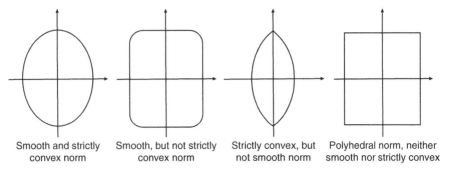

| Smooth and strictly convex norm | Smooth, but not strictly convex norm | Strictly convex, but not smooth norm | Polyhedral norm, neither smooth nor strictly convex |

Fig. 1.2 Comparison of smooth, strictly convex, and polyhedral norms

In the following, a result from Convex Analysis is stated in order to define the *polar* or *dual* of a norm. Afterwards, properties of smooth, strictly convex, and polyhedral norms are discussed. Finally, the *bisector* between two points is defined and properties depending on the underlying norm are discussed.

1.3.1 Dual Norms

We define the *support function* and the *polar* of a convex set $C \subseteq \mathbb{R}^n$.

Definition 1.19 ([Roc70]). The *support function* of a convex set $C \subseteq \mathbb{R}^n$ is defined by

$$\delta^*(X|C) = \sup\{X^t Y : Y \in C\}.$$

Definition 1.20 ([Roc70]). Let $C \subseteq \mathbb{R}^n$ be nonempty and convex. Then the *polar set* of C is given by

$$C^\circ = \{X \in \mathbb{R}^n : X^t Y \le 1 \ \forall Y \in C\}.$$

We have the following relation between a closed convex set and the corresponding polar set.

Theorem 1.21 ([Roc70]). *Let C be a closed convex set containing the origin. The polar C° is then another closed convex set containing the origin, and $C^{\circ\circ} = C$. If C is additionally symmetric with respect to the origin, then C° inherits this property. The norm $\|\cdot\|$ with unit ball C is the support function of C°; that is,*

$$\|X\| = \delta^*(X|C^\circ).$$

Dually, the norm with unit ball C° is the support function of C.

This result justifies the following definition.

Definition 1.22. Let $\| \cdot \|$ be a norm with unit ball B. Then the norm $\| \cdot \|^\circ$ with unit ball B° is called *dual norm* or *polar norm* of $\| \cdot \|$.

Remark 1.23. Note that Rockafellar [Roc70] mainly considered the more general case of *gauge functions*. A gauge function γ_B is defined according to (1.4) but without any restriction on the symmetry of B. Theorem 1.21 is still valid if *norm* is replaced by *gauge function*.

1.3.2 Smooth Norms

For smooth norms a lot of different characterizations are known. We only consider a few geometric characterizations. We start with a result for smooth norms in the real plane $\mathbb{R}^2$ which goes back to Kramer and Németh [KN73].

Theorem 1.24. *A norm $\| \cdot \|$ on $\mathbb{R}^2$ is smooth if and only if for any triple of three noncollinear points A_1, A_2, A_3 there exists at least one point $X \in \mathbb{R}^2$ such that*

$$\|A_1 - X\| = \|A_2 - X\| = \|A_3 - X\|.$$

Proof. See [KN73], or alternatively [MSW01]. $\qquad\square$

The forward implication of Theorem 1.24 also applies to higher dimensions.

Theorem 1.25 ([Gro69]). *If $\| \cdot \|$ is a smooth norm in $\mathbb{R}^n$, then for any $n+1$ points $A_1, \ldots, A_{n+1} \in \mathbb{R}^n$ in general position there exists at least one point $X \in \mathbb{R}^n$ such that*

$$\|A_1 - X\| = \|A_2 - X\| = \ldots = \|A_{n+1} - X\|.$$

Using optimal solutions to the hyperplane location problem with minisum objective function a further characterization of smooth norms is possible. In order to state this characterization define

$$f(H) := \sum_{m=1}^{M} \omega_m \min\{\|A_m - Y\| : Y \in H\}$$

where $H \subseteq \mathbb{R}^n$ is a hyperplane.

Theorem 1.26 ([MS99]). *Let $\| \cdot \|$ be a norm in $\mathbb{R}^n$. $\| \cdot \|$ is smooth if and only if for all instances of the hyperplane location problem with minisum objective function $f(H)$ all optimal hyperplanes are affine hulls of n affinely independent fixed points.*

Remark 1.27. A characterization of smooth norms similar to Theorem 1.26 is also known for hyperplane location problems with minimax objective function, see [MS99].

1.3.3 Strictly Convex Norms

Also for strictly convex norms a lot of different characterizations are known. As for smooth norms, we mainly consider geometric characterizations.

Theorem 1.28 (e.g., [Haz90]). *Let $\mathscr{X}$ be a real Banach space with norm $\|\cdot\|$. The following statements are equivalent to strictly convexity of $\|\cdot\|$:*

1. For $X \neq Y \in \mathscr{X}$, $\|X\| = \|Y\| = 1$, we have

$$\|\lambda X + (1 - \lambda)Y\| < 1 \text{ for all } \lambda \in]0,1[.$$

2. If $X \neq Y \in \mathscr{X}$ and $\|X\| = \|Y\| = 1$, then $\|X + Y\| < 2$.
3. If for $X, Y, Z \in \mathscr{X}$ we have

$$\|X - Y\| = \|X - Z\| + \|Z - Y\|,$$

then there exists $\lambda \in [0,1]$ such that $Z = \lambda X + (1 - \lambda)Y$.
4. Any $X^* \in \mathscr{X}^*$ has at most one maximum on the unit ball.

Proof. See [Haz90, Chapter 2].   □

Many further characterizations are listed in the survey on Minkowski spaces given by Martini et al. [MSW01, MS04]. We give a small selection of these results.

Theorem 1.29 ([MSW01]). *Let $\mathscr{X}$ be a finite dimensional real Banach space with norm $\|\cdot\|$. The following statements are equivalent to strictly convexity of $\|\cdot\|$:*

1. Every boundary point of the unit ball is an extreme point (exposed point).
2. The unit ball is rotund.
3. Any supporting hyperplane of the unit ball touches the unit ball in at most one boundary point.
4. Supporting hyperplanes at distinct points of the boundary of the unit ball are distinct.
5. For each point X and each convex set $D \subseteq \mathscr{X}$ there is at most one point in D that is nearest to X.
6. For each point X and each closed convex set $D \subseteq \mathscr{X}$ there is exactly one point in D that is nearest to X.

Also a result related to Theorem 1.24 applies to strictly convex norms.

Theorem 1.30. *A norm $\|\cdot\|$ on $\mathbb{R}^2$ is strictly convex if and only if for any triple of three points A_1, A_2, A_3 there exists at most one point $X \in \mathbb{R}^2$ such that*

$$\|A_1 - X\| = \|A_2 - X\| = \|A_3 - X\|.$$

Proof. See Prop. 14 on p. 106 in [MSW01].   □

Combining Theorems 1.24 and 1.30 it can be concluded that a norm $\|\cdot\|$ in $\mathbb{R}^2$ is smooth and strictly convex if and only if for any triple of three points A_1, A_2, A_3 there exists *exactly one* point $X \in \mathbb{R}^2$ such that

$$\|A_1 - X\| = \|A_2 - X\| = \|A_3 - X\|.$$

1.3.4 Polyhedral Norms

According to Definition 1.17 the unit ball B of a polyhedral norm is a full-dimensional convex polytope which is symmetric with respect to the origin. Since such a polytope has a finite number of extreme points, we may denote in the following the extreme points of B as

$$\text{Ext}(B) = \{E_g : 1 \leq g \leq s\}.$$

Since B is symmetric with respect to the origin, $s \in \mathbb{N}$ is always an even number and for any $E_i \in \text{Ext}(B)$ there exists a counterpart $E_j \in \text{Ext}(B)$ such that $E_i = -E_j$.

Notation 1.31. Let $\|\cdot\|$ be a polyhedral norm and let $B \subseteq \mathbb{R}^n$ be its unit ball. Then the extreme points $\text{Ext}(B)$ of B are also called *fundamental directions* of the polyhedral norm $\|\cdot\|$.

Notation 1.32. Let $\|\cdot\|$ be a polyhedral norm. Any straight line parallel to a fundamental direction of $\|\cdot\|$ is called *direction line* of the polyhedral norm $\|\cdot\|$.

By increasing the number of fundamental directions, a polyhedral norm can become as accurate a representation of an arbitrary norm as desired. Thus, we have the following result.

Theorem 1.33 ([WWR85]). *The class of polyhedral norms is a dense subset of all norms in $\mathbb{R}^n$.*

The following theorem states a characterization of polyhedral norms.

Theorem 1.34 ([WWR85]). *A polyhedral norm $\|\cdot\|$ has a characterization as*

$$\|X\| = \min\left\{\sum_{g=1}^{s} |\beta_g| : X = \sum_{g=1}^{s} \beta_g E_g\right\}. \tag{1.6}$$

As mentioned above, it is assumed that each fundamental direction E_i of the polyhedral norm $\|\cdot\|$ has a counterpart E_j such that $E_i = -E_j$. Therefore, we have the following linear formulation.

Corollary 1.35. *For any $X \in \mathbb{R}^n$ we have*

$$\|X\| = \min \left\{ \sum_{g=1}^{s} \beta_g : X = \sum_{g=1}^{s} \beta_g E_g, \ \beta_g \geq 0 \ \forall \ g = 1, \ldots, s \right\}.$$

Remark 1.36. Historically, our definition of polyhedral norms (see Definition 1.17) is not correct. Witzgall [Wit64] defined polyhedral norms as the class of norms having the characterization given by (1.6). Ward and Wendell [WWR85] then defined *block norms* as the class of norms having a polyhedral unit ball. Theorem 1.34 shows that both definitions are equivalent.

In order to state our next result we need the following notation.

Notation 1.37. Let $\| \cdot \|$ be a polyhedral norm in $\mathbb{R}^2$ with fundamental directions $\text{Ext}(B) = \{E_g : 1 \leq g \leq s\}$. For any $E_g \in \text{Ext}(B)$ we denote as E_g^+ the fundamental direction adjacent to E_g in clockwise direction.

Note that $E_g^+ = E_{g+1}$ for $1 \leq g \leq s - 1$, provided that the fundamental directions $\text{Ext}(B) = \{E_g : 1 \leq g \leq s\}$ of a polyhedral norm in $\mathbb{R}^2$ are ordered in clockwise direction. Furthermore, in this case we have $E_s^+ = E_1$. For any index g, $1 \leq g \leq s$, we define Γ_g as the cone generated by E_g and E_g^+. The following lemma shows that a polyhedral norm $\| \cdot \|$ in $\mathbb{R}^2$ can be easily evaluated in a point X, provided that $X \in \Gamma_g$.

Lemma 1.38 ([NP05]). *Let $\| \cdot \|$ be a polyhedral norm in $\mathbb{R}^2$ with fundamental directions $\text{Ext}(B) = \{E_g : 1 \leq g \leq s\}$. Furthermore, let $X \in \Gamma_g$ such that $X = \beta_g E_g + \beta_{g^+} E_g^+$. Then $\|X\| = \beta_g + \beta_{g^+}$.*

Lemma 1.38 can be generalized to higher dimensions. To this end, we need to generalize the cones Γ_g.

Notation 1.39. Given a polyhedral norm $\| \cdot \|$ in $\mathbb{R}^n$ with unit ball $B \subseteq \mathbb{R}^n$ and a facet $\mathscr{F}$ of B, let $\text{Ext}(\mathscr{F}) = \{E_1, \ldots, E_t\}$ denote the extreme points of $\mathscr{F}$ and let

$$\Gamma_{\mathscr{F}} := \left\{ X \in \mathbb{R}^n : X = \sum_{g=1}^{t} \beta_g E_g, \ \beta_g \geq 0, \ E_g \in \text{Ext}(\mathscr{F}) \right\}.$$

$\Gamma_{\mathscr{F}}$ is called *fundamental cone* of the polyhedral norm $\| \cdot \|$.

Lemma 1.40 ([SKW02]). *Let $\| \cdot \|$ be a polyhedral norm in $\mathbb{R}^n$ with unit ball $B \subseteq \mathbb{R}^n$. Let $\overline{X} \in \Gamma_{\mathscr{F}}$ where $\Gamma_{\mathscr{F}}$ is the fundamental cone defined by the extreme points $\text{Ext}(\mathscr{F}) = \{E_1, \ldots, E_t\}$, $t \geq n$. Let $\overline{X} = \sum_{g=1}^{t} \beta_g E_g$ be a representation of $\overline{X}$ in terms of $E_1, \ldots, E_t$. Then $\|\overline{X}\| = \sum_{g=1}^{t} \beta_g$.*

Remark 1.41. Note that the representation $\overline{X} = \sum_{g=1}^{t} \beta_g E_g$ is not unique in general. However, all representations $\overline{X} = \sum_{g=1}^{t} \beta_g E_g$ can be used to calculate $\|\overline{X}\|$, even representations where one or more β_g are negative (cf. [SKW02]).

Polyhedral norms in $\mathbb{R}^2$ may also be characterized in the following way.

Theorem 1.42 ([WWR85]). *A norm* $\|\cdot\|$ *is a polyhedral norm in* $\mathbb{R}^2$ *if and only if there exists a subset* $\{V_1, \ldots, V_t\} \subseteq \mathbb{R}^2$ *that spans* $\mathbb{R}^2$ *such that*

$$\|X\| = \sum_{g=1}^{t} |X^t V_g| \quad \text{for all } X \in \mathbb{R}^2. \tag{1.7}$$

Remark 1.43. A norm in $\mathbb{R}^n$ that has a representation according to (1.7) is called *additive norm*, see [WWR85]. Thus Theorem 1.42 shows that the class of additive norms and the class of polyhedral norms in $\mathbb{R}^2$ coincide.

Finally, we consider the dual of a polyhedral norm.

Theorem 1.44 ([WWR85]). *Let* $\|\cdot\|$ *be a polyhedral norm. Then its dual is also a polyhedral norm. Furthermore, a polyhedral norm in* $\mathbb{R}^2$ *and its dual have the same number of fundamental directions, i.e.,* $|\text{Ext}(B)| = |\text{Ext}(B^\circ)|$.

Since the unit ball B° of the dual of a polyhedral norm is a polytope, its support function is given as

$$\delta^*(X|B^\circ) = \max\{|X^t E_g^\circ| : E_g^\circ \in \text{Ext}(B^\circ)\}.$$

Therefore, we obtain as a consequence of Theorem 1.21 the following result.

Corollary 1.45. *Let* $\|\cdot\|$ *be a polyhedral norm with unit ball B. Then*

$$\|X\| = \max\{|X^t E_g^\circ| : E_g^\circ \in \text{Ext}(B^\circ)\}.$$

1.3.5 Bisectors

A set which strongly depends on the classes of norms is the bisector of two points. In the following we summarize a few results on bisectors under smooth, strictly convex, and polyhedral norms. We start with a definition of bisectors.

Definition 1.46. Let $\mathscr{X}$ be a Banach space with norm $\|\cdot\|$ and let $A_1, A_2 \in \mathscr{X}$. The *bisector* of A_1 and A_2 is defined as

$$B(A_1, A_2) = \{X \in \mathscr{X} : \|A_1 - X\| = \|A_2 - X\|\}.$$

In the Euclidean case the bisector of two points is simply the well-known perpendicular bisector. In particular, the Euclidean bisector is a hyperplane. In general, a bisector of two points in $\mathbb{R}^n$ is not a hyperplane; even more, it may contain n-dimensional subsets as the following example shows.

Example 1.47. In Fig. 1.3 two bisectors of the Manhattan norm in the real plane $\mathbb{R}^2$ are illustrated. In Fig. 1.3a the line segment joining A_i and A_j is not parallel to an edge of the unit ball of the Manhattan norm; in Fig. 1.3b the line segment is parallel to such an edge. The first condition yields a piecewise linear curve for $B(A_i, A_j)$,

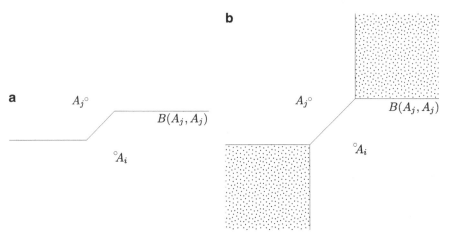

Fig. 1.3 Bisectors of the Manhattan norm for different conditions. (**a**) Bisector is piecewise linear curve. (**b**) Bisector contains subregions

while the second condition leads to a piecewise linear curve joining two symmetric subregions (or cones) bounded by adjacent fundamental directions of the Manhattan norm.

The results of Example 1.47 may be generalized to arbitrary norms. To this end we need the topological concept of *homeomorphic* sets, see for instance [GG99].

Theorem 1.48 (e.g., [MS04, Ma00]). *Let $\mathcal{X}$ be a two-dimensional real Banach space with norm $\|\cdot\|$ and $A_1 \neq A_2 \in \mathcal{X}$. Then the bisector $B(A_1, A_2)$ is homeomorphic to a line if and only if the straight line $\ell_{12} = \langle A_1, A_2 \rangle$ is not parallel to a flat spot of the boundary of the unit ball of $\|\cdot\|$. $B(A_1, A_2)$ contains two two-dimensional regions if and only if ℓ_{12} is parallel to a flat spot of the boundary of the unit ball of $\|\cdot\|$.*

Since the boundary of the unit ball of a strictly convex norm does not contain any flat spot, Theorem 1.48 implies the following corollary.

Corollary 1.49. *The norm of a two-dimensional real Banach space is strictly convex if and only if any bisector is homeomorphic to a straight line.*

The forward implication of Corollary 1.49 also applies to higher dimensions, the backward implication is false in any dimension $n \geq 3$. This is shown by the following theorem which is stated in this form in [MS04].

Theorem 1.50 ([MS04]). *On the one hand, all bisectors are homeomorphic to a hyperplane if the norm of a finite dimensional real Banach space is strictly convex. On the other hand, for any integer $n \geq 3$ there exists an n-dimensional real Banach space with nonstrictly convex norm $\|\cdot\|$ such that each bisector is homeomorphic to a hyperplane.*

The proof of the forward implication of Theorem 1.50 goes back to a result of Horváth given in [Hor00]; the backward implication is shown by an example which also goes back to Horváth.

Theorem 1.50 gives a first hint on the differences between bisectors in two-dimensional space and bisectors in higher dimensions. A detailed discussion of the two-dimensional case vs. the three-dimensional case may be found for instance in the paper by Icking et al. [IKLM94]. However, also bisectors in two-dimensional spaces can behave unexpectedly.

Theorem 1.51 ([CMR96, MS04]). *In a two-dimensional real Banach space pairs of bisectors can exist that intersect infinitely many times.*

As mentioned above, bisectors defined with respect to the Euclidean norm in $\mathbb{R}^n$ are hyperplanes. But do there exist other norms having this property? An answer to this important question goes back to Day.

Theorem 1.52 ([Day47]). *All bisectors in a finite-dimensional real Banach space with norm $\| \cdot \|$ are hyperplanes if and only if the unit ball of $\| \cdot \|$ is an ellipsoid.*

Definition 1.53. The norm $\| \cdot \|$ of a finite-dimensional real Banach space is called *elliptic* norm if its unit ball is an ellipsoid.

We close our short survey on bisectors with a result on the complexity of a bisector defined with respect to a polyhedral norm in $\mathbb{R}^2$.

Theorem 1.54 ([NP05]). *Let $n = 2$ and $\| \cdot \|$ a polyhedral norm in $\mathbb{R}^2$ with s fundamental directions. Given two points $X, Y \in \mathbb{R}^2$ the bisector $B(X, Y)$ consists of at most $4s$ different subsets defined by different linear equations.*

1.4 Finite Dominating Sets

In this section we give a formal definition of the notion *finite dominating set* which is frequently used in Operations Research-related literature. As far as the author is aware of, finite dominating sets are not defined in any prior publication in a strict manner.

We start with a formal definition of an optimization problem (or mathematical programming problem).

Definition 1.55 (Optimization problem). An *optimization problem* is a quadruple $P = (\mathscr{I}, \mathscr{F}, f, \text{opt})$ such that

- $\mathscr{I}$ is the set of *input instances* for P.
- $\mathscr{F}(I)$ is the set of *feasible solutions* for input $I \in \mathscr{I}$.
- $f : \{(I, S) : S \in \mathscr{F}(I), I \in \mathscr{I}\} \to \mathbb{R}$.
- $\text{opt} \in \{\inf, \sup\}$.

Given an instance I of an optimization problem $P = (\mathcal{I}, \mathcal{F}, f, \text{opt})$ the goal is to find a feasible solution $X \in \mathcal{F}(I)$ such that

$$f(I, X) = \text{opt}\{f(I, Y) : Y \in \mathcal{F}(I)\}.$$

Definition 1.56. Let $P = (\mathcal{I}, \mathcal{F}, f, \text{opt})$ be an optimization problem. For any $I \in \mathcal{I}$ we define

$$\text{opt}(I) := \text{opt}\{f(I, Y) : Y \in \mathcal{F}(I)\},$$
$$\mathcal{F}(I)^* := \{Y \in F(I) : f(I, Y) = \text{opt}(I)\}.$$

Given an optimization problem we may define a finite dominating set in the following way.

Definition 1.57 (Finite dominating set). Let $P = (\mathcal{I}, \mathcal{F}, f, \text{opt})$ be an optimization problem. Each function

$$\Xi : \mathcal{I} \to \bigcup_{I \in \mathcal{I}} \mathcal{F}(I)$$

is called *finite dominating set* for P if the intersection $\Xi(I) \cap \mathcal{F}(I)^*$ is nonempty and finite; that is, $0 < |\Xi(I) \cap \mathcal{F}(I)^*| < \infty$ for all $I \in \mathcal{I}$. We also use the abbreviation *FDS*.

A finite dominating set Ξ is a mapping that assigns to each instance $I \in \mathcal{I}$ a finite set $\Xi(I)$ containing at least one optimal solution $Y \in \mathcal{F}(I)^*$. Note that for any optimization problem there exists a finite dominating set, provided that $\mathcal{F}(I) \neq \emptyset$ for all $I \in \mathcal{I}$. Indeed, let $P = (\mathcal{I}, \mathcal{F}, f, \text{opt})$ be an optimization problem such that $\mathcal{F}(I) \neq \emptyset$ for all $I \in \mathcal{I}$ and let $Y_I \in \mathcal{F}_I$ for all $I \in \mathcal{I}$. Then $\Xi(I) := \{Y_I\}$ is an FDS according to Definition 1.57.

Remark 1.58. Note that Definition 1.57 does not make a statement on the *computability* of a finite dominating set Ξ. Even though in almost all cases a finite dominating set for an optimization problem exists, it is possible that a computable FDS does not exist. For instance, Bajaj proved in [Baj88] that an exact algorithm for the (Euclidean) Weber problem cannot exist under models of computation where the root of an algebraic equation is obtained using arithmetic operations and the extraction of kth roots. A consequence of this result is that no computable finite dominating set can exist for this problem.

The most famous example for a finite dominating set is the fundamental theorem of linear programming, see, e.g., [Lue04]. Basically, this theorem says that given a linear program $P = (\mathcal{I}, \mathcal{F}, f, \text{opt})$ the function Ξ which maps each input I on the extreme points of $\mathcal{F}(I)$ is an FDS. The simplex method is based on this FDS.

An interesting question is the following:

Given an optimization problem $P = (\mathcal{I}, \mathcal{F}, f, \text{opt})$ does there exist an FDS Ξ for P and an efficient (polynomial) algorithm that computes the set $\Xi(I)$ for all $I \in \mathcal{I}$?

If the optimization problem P corresponds to a decision problem P_d in the complexity class P [GJ79], then the answer is yes: $\Xi(I)$ may be defined as a set containing exactly one optimal solution to P and the computation of the elements of $\Xi(I)$ may be done efficiently by the polynomial algorithm for P. If a problem belongs to the complexity class NP but not to P, then it may be possible that the answer is also positive. But in this case it cannot be possible to evaluate each point of the FDS efficiently. Otherwise, it follows that the problem belongs to the complexity class P.

Independent from the complexity of a problem $P = (\mathscr{I}, \mathscr{F}, f, \text{opt})$, it may happen that a finite dominating set Ξ for P cannot be evaluated in each point efficiently. Indeed, suppose that the size of the sets $\Xi(I)$, $I \in \mathscr{I}$, is not bounded by some polynomial in terms of the input length of the problem. Then the brute force algorithm which tests all solutions $Y \in \Xi(I)$ is not efficient. An example for this situation is the fundamental theorem of linear programming. Nevertheless, even if the size of an FDS Ξ is not bounded by some polynomial, the knowledge of Ξ and an efficient algorithm to compute $\Xi(I)$ may be very useful in order to obtain good algorithms for optimization problems. For instance, an FDS may convert a continuous optimization problem into a discrete problem. Therefore an FDS allows to tackle continuous problems with tools of discrete optimization. This approach is in particular interesting for global optimization problems with various local optima. Numerous examples of solution methods based on finite dominating sets for global optimization problems may be found in the literature, see for instance [HD04, NP05, Nic95] and the references within these books.

Chapter 2
Euclidean Minisum Hyperspheres

2.1 Basic Assumptions

In this chapter the Euclidean minisum hypersphere problem is studied which may be described as follows:

Dimensions: $n \geq 2, M \geq 2$
Norm: Euclidean norm $\|\cdot\| := \|\cdot\|_2$ in $\mathbb{R}^n$
Objects: Hyperspheres defined with respect to $\|\cdot\|_2$
Metric: Induced by $\|\cdot\|_2$; $d(X,Y) := \|Y - X\|_2$
Distance: $d(S,A_m) = \min\{d(Y,A_m) : Y \in S\}, 1 \leq m \leq M$
Objective: $f(S(X,r)) = \sum_{m=1}^{M} \omega_m d_m(S(X,r)) \to \min$

The Euclidean minisum hypersphere problem is discussed by Brimberg et al. [BJS09b] for the two-dimensional case and by Nievergelt [Nie10] for arbitrary dimensions ($n \geq 1$). In this chapter their results are summarized and some alternative proofs are given. Also new results are stated and counterexamples to conjectures are given. Throughout this chapter $M \geq n+2$ is assumed. Furthermore, in this chapter a hypersphere is always defined with respect to the Euclidean norm. For the sake of simplicity $\|\cdot\|$ refers to the Euclidean norm, i.e., for $X = (x_1, \ldots, x_n) \in \mathbb{R}^n$

$$\|X\| := \|X\|_2 = \sqrt{\sum_{i=1}^{n} x_i^2}.$$

M.-C. Körner, *Minisum Hyperspheres*, Springer Optimization and Its Applications 51, DOI 10.1007/978-1-4419-9807-1_2, © Springer Science+Business Media, LLC 2011

2.2 Distance

The Euclidean point–hypersphere distance can be readily obtained: Given a hypersphere $S(X,r) \in \mathscr{G}$ and a point $A \in \mathbb{R}^n$, the point–hypersphere distance between $S(X,r)$ and A is given as the absolute value of the difference of the radius r and $\|X - A\|$. This result is shown in the following lemma.

Lemma 2.1. *Let $A \in \mathbb{R}^n$ be a point and let $S(X,r) \in \mathscr{G}$ be a hypersphere with center X and radius r. Then we have*

$$d(S(X,r),A) = |\|X - A\| - r|.$$

Proof. Let $P = S(X,r) \cap [X,A\rangle$. Since the points X, A, and P are collinear, the triangle inequality holds with equality in each of the following three cases (cf. Lemma 1.13 and Corollary 1.15):

- If A is outside the hypersphere, then P lies between A and X. We obtain

$$\|X - A\| = \|X - P\| + \|P - A\| = r + \|P - A\|. \tag{2.1}$$

- If A is inside the hypersphere, then A lies between P and X. We obtain

$$\|X - A\| = \|X - P\| - \|P - A\| = r - \|P - A\|. \tag{2.2}$$

- If A lies on the hypersphere, then $A = P$ and we obtain

$$\|X - A\| = \|X - P\| = r. \tag{2.3}$$

In all cases we get that $\qquad \|P - A\| = |\|X - A\| - r|.$

Now, we show $\|P - A\| \le d(S(X,r),A)$. To this end we choose an arbitrary point $Y \in S(X,r)$ and show $\|P - A\| \le \|Y - A\|$. Again, we distinguish three cases:

1. If A is outside the hypersphere the triangle inequality and (2.1) imply

$$\|Y - A\| \ge \|X - A\| - \|X - Y\| = \|X - A\| - r = \|P - A\|.$$

2. If A is inside the hypersphere the triangle inequality and (2.2) imply

$$\|Y - A\| \ge \|Y - X\| - \|A - X\| = r - \|X - A\| = \|P - A\|.$$

3. If A lies on the hypersphere (2.3) implies

$$\|Y - A\| \ge 0 = \|P - A\|. \qquad \square$$

Using Lemma 2.1 the minisum hypersphere problem may be rewritten as

$$f(S(X,r)) = \sum_{m=1}^{M} \omega_m |\|A - X\| - r| \to \min.$$

Beyond that, we can draw some further consequences.

Corollary 2.2. *Let $A, X \in \mathbb{R}^n$. Then the point–hypersphere distance $d(S(X,r),A)$ is convex and piecewise linear in r.*

Corollary 2.3. *Let $X, A \in \mathbb{R}^n$. Then the distance $d(S(X,r),A)$ is*

(i) concave in (X,r) on the set

$$\mathcal{V} = \{(X,r) \in \mathbb{R}^n \times]0, \infty[: \|X - A\| \leq r\}$$

and
(ii) convex in (X,r) on any convex set $U \subseteq (\mathbb{R}^n \times]0, \infty[) \setminus \mathcal{V}$.

Proof. Follows immediately from the convexity of $\| \cdot \|$ and Lemma 2.1. $\qquad \square$

Roughly speaking, Corollary 2.3 says that $d(S(X,r),A)$ behaves like a convex function if A stays outside the hypersphere $S(X,r)$ while moving X and r. If A stays inside $S(X,r)$ for small movements of X and r, then $d(S(X,r),A)$ behaves like a concave function.

The point–hypersphere distance has the following symmetry property.

Corollary 2.4. *Let $A \in \mathbb{R}^n$ and $S(X,r) \in \mathcal{G}$. Then we have $S(A,r) \in \mathcal{G}$ and*

$$d(S(X,r),A) = d(S(A,r),X).$$

Proof. $S(A,r) \in \mathcal{G}$ is clear. Therefore $d(S(A,r),X)$ is well defined and the result follows from Lemma 2.1. $\qquad \square$

2.3 Degenerated Solutions

A point may be interpreted as a degenerated hypersphere with radius $r = 0$. Analogously, a hyperplane may be interpreted as a degenerated hypersphere with radius $r = \infty$. It is easy to see that a minisum hypersphere cannot be degenerated to a point.

Lemma 2.5. *A hypersphere $S(X,r)$ with radius $r = 0$ cannot be a minisum hypersphere.*

Proof. Assume that $S_0 = S(X_0, 0)$ is a minisum hypersphere. Since we assumed $M \geq 2$, we can find a fixed point, say A_1, such that $X_0 \neq A_1$. It follows that $\|X_0 - A_1\| > 0$. Using Theorem 1.25 it follows that there exists a hypersphere S_1

intersecting X_0 and A_1. We compare the objective values of S_0 and S_1: Note that $S_0 \subset S_1$; hence $d(S_0, A_m) \geq d(S_1, A_m)$ for all $m = 1, \ldots, M$. In particular for $m = 1$ we obtain a strict inequality, namely,

$$d(S_0, A_1) = \|X_0 - A_1\| > 0 = \|A_1 - A_1\| = d(S_1, A_1).$$

Together

$$f(S_0) = \sum_{m=1}^{M} \omega_m d(S_0, A_m) > \sum_{m=1}^{M} \omega_m d(S_1, A_m) = f(S_1).$$

Thus S_0 cannot be optimal.   □

The proof of Lemma 2.5 is a generalization of the proof given by Brimberg et al. [BJKS09]. Note that the proof of Lemma 2.5 is still valid if the Euclidean norm $\| \cdot \|$ is replaced by an arbitrary norm $\| \cdot \|$. A result similar to Lemma 2.5 may be found in [Nie10].

Whereas a point cannot be a minisum hypersphere, it is possible that the objective value of a hyperplane is superior to the objective value of a minisum hypersphere. To see this, we need the following definition.

Definition 2.6. The set of all hyperspheres and hyperplanes in $\mathbb{R}^n$ is denoted as $\overline{\mathscr{G}}$. The elements of $\overline{\mathscr{G}}$ are called *generalized hyperspheres*, see [Nie10, Ber87].

Notation 2.7. We denote the minisum hypersphere problem with feasible set $\overline{\mathscr{G}}$ as *generalized minisum hypersphere problem*. An optimal solution to this problem is called *generalized minisum hypersphere*.

We have the following result.

Lemma 2.8. *For each $n \geq 2$ there exists a finite set $\mathscr{D} \subseteq \mathbb{R}^n$ such that no optimal solution $S \in \overline{\mathscr{G}}$ to the generalized minisum hypersphere problem is a hypersphere. For the minisum hypersphere problem this means that in this case no minisum hypersphere $S \in \mathscr{G}$ exists.*

Proof. Let $\mathscr{D} \subseteq \mathbb{R}^n$ be a set of three distinct points lying on a common straight line. From Theorem 1.29 we may conclude that there does not exist any hypersphere $S \in \mathscr{G}$ intersecting all points in $\mathscr{D}$. Thus, $f(S) > 0$ for all hyperspheres $S \in \mathscr{G}$.

By contrast, consider any hyperplane $H \in \overline{\mathscr{G}}$ with $\mathscr{D} \subseteq H$. The distance between each point in $\mathscr{D}$ and the hyperplane H is zero. Thus, we have $f(H) = 0$ and the minimum value of f on $\overline{\mathscr{G}}$ is zero.

Since it is clear that a sequence of hyperspheres exists which converges against H, zero is the infimum but not the minimum value of f on $\mathscr{G}$. Hence, $\mathscr{G}$ does not contain an optimal solution.   □

An alternative proof of Lemma 2.8 may be found in [Nie10].

However, $M \geq n + 3$ and no collinear triple in $\mathscr{D}$ ensure that no optimal solution to the generalized minisum hypersphere problem can be a hyperplane. This result is a generalization of an analog result stated in [BJS09b] for the planar case.

Theorem 2.9. *Let $\mathscr{D} \subseteq \mathbb{R}^n$ be in general position. Suppose $M \geq n+3$ and that no triple of the fixed points is collinear. Then no optimal solution to the generalized minisum hypersphere problem is a hyperplane.*

We postpone the proof of Theorem 2.9 to Section 3.3. There we extend Theorem 2.9 to elliptic norms and prove the result in a more general setting.

We close this section with an example showing that $M \geq n+3$ in Theorem 2.9 is necessary.

Example 2.10. Let $n = 2$, $M = 4$; assume a configuration of the fixed points as depicted in Fig. 2.1, and let $\omega_1 = \omega_2 = 100$, $\omega_3 = \omega_4 = 1$. Then any circle passing through A_1 and A_2 and intersecting the line segment $[A_3, A_4]$ has objective value 2 and is a minisum circle. But also the straight line $\langle A_2, A_4 \rangle$ has objective value 2 and therefore it is an optimal solution to the generalized minisum circle problem.

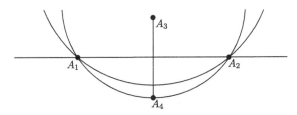

Fig. 2.1 Illustration of Example 2.10

2.4 Existence of Optimal Solutions

If the set of fixed points $\mathscr{D}$ does not satisfy the conditions of Theorem 2.9, then a hyperplane could be superior to all hyperspheres. In this case no optimal solution to the minisum hypersphere problem exists. The goal of this section is to show that for the generalized minisum hypersphere problem always an optimal solution exists; that is, there exists either a hypersphere or a hyperplane S^* such that

$$\inf\{f(S) : S \in \overline{\mathscr{G}}\} = f(S^*).$$

Having this result, we can draw the following conclusion: If a hyperplane is not an optimal solution to the generalized minisum hypersphere problem, then a minisum hypersphere exists. Hence, a minisum hypersphere exists if the conditions of Theorem 2.9 are met by the fixed points in $\mathscr{D}$.

First, we note that any minisum hypersphere must intersect the convex hull of the fixed points.

Lemma 2.11 ([Nie10]). *If a hypersphere S_1 does not intersect the convex hull of the fixed points $\mathscr{D}$, then there exists a hypersphere S_2 superior to S_1.*

Proof. See [Nie10] for the Euclidean case. In Corollary 3.20 we state a similar result for the more general case of arbitrary norms.   □

Now, we study the set $\overline{\mathscr{G}}$ in more detail. To this end consider the equation

$$0 = u\|Y\|^2 + 2V^t Y + w \tag{2.4}$$

where $u, v \in \mathbb{R}$, $Y, V \in \mathbb{R}^n$, see [Ber87]. For $u \neq 0$ this equation is equivalent to

$$0 = \frac{1}{u} \left(\|uY + V\|^2 + uw - \|V\|^2 \right). \tag{2.5}$$

Equations (2.4) and (2.5) induce a correspondence between points $(u, V, w) \in \mathbb{R} \times \mathbb{R}^n \times \mathbb{R}$ and generalized hyperspheres. Indeed, for any point $(u, V, w) \in \{0\} \times (\mathbb{R}^n \setminus \{0\}) \times \mathbb{R}$, (2.4) describes the hyperplane

$$H = \left\{ Y \in \mathbb{R}^n : V^t Y + 2^{-1} w = 0 \right\};$$

for $(u, V, w) \in (\mathbb{R} \setminus \{0\}) \times \mathbb{R}^n \times \mathbb{R}$ we have

$$\left\{ Y \in \mathbb{R}^n : \frac{1}{u} \left(\|uY + V\|^2 + uw - \|V\|^2 \right) = 0 \right\}$$
$$= \left\{ Y \in \mathbb{R}^n : u\|Y + u^{-1}V\|^2 = u^{-1}(\|V\| - uw) \right\}$$
$$= \left\{ Y \in \mathbb{R}^n : \|Y - (-u^{-1}V)\| = \sqrt{u^{-2}(\|V\|^2 - uw)} \right\}.$$

Hence, (2.5) describes the hypersphere $S(X, r)$ where

$$X = -u^{-1}V,$$
$$r = \sqrt{u^{-2}(\|V\|^2 - uw)}.$$

Thus, we can parameterize the set $\overline{\mathscr{G}}$ by $\mathbb{R}^{n+2}$. Not all points (u, V, w) correspond to generalized hyperspheres. For instance, we have $S(X, r) = \emptyset$ whenever $\|V\|^2 - uw < 0$. Defining $\overline{\mathscr{G}}^*$ as the set of all points $(u, V, w) \in \mathbb{R}^{n+2}$ where $\|V\|^2 - uw \geq 0$ holds, Nievergelt [Nie10] was able to prove the following result.

Lemma 2.12 ([Nie10]). *For all $n \geq 2$ and any point $A \in \mathbb{R}^n$, the distance $d(S, A)$ between A and the generalized hypersphere with coefficients (u, V, w) is continuous on $\overline{\mathscr{G}}^* \setminus \{(0, 0, 1)\}$ and diverges to infinity as (u, V, w) tends to $(0, 0, 1)$.*

The proof of Lemma 2.12 is mainly based on results from Algebraic Geometry. We do not present it here and refer the reader to the original work of Nievergelt.

Combining Lemmas 2.11 and 2.12 we obtain the main result of this section.

Theorem 2.13 ([Nie10]). *There always exists an optimal solution to the generalized minisum hypersphere problem.*

Proof. Let $S(X, r)$ be any (not necessary optimal) solution to the generalized minisum hypersphere problem. Then for each fixed point A_m consider the set L_m of points $(u, V, w) \in \overline{\mathscr{G}}^*$ that correspond to generalized hyperspheres having a distance to A_m which is less than or equal to $d_m(S(X, r))$. By using Lemma 2.12 we may conclude that L_m is compact for each $1 \leq m \leq M$. Hence the intersection $L = \bigcap_{1 \leq m \leq M} L_m$ is also compact. Considering the objective function f as a function on $\overline{\mathscr{G}}^*$ it follows that f attains a minimum on L. Using Lemma 2.11 it follows that this minimum is a global minimum of f on $\overline{\mathscr{G}}^*$.   □

Corollary 2.14. *Let $\mathscr{D} \subseteq \mathbb{R}^n$ be in general position. Suppose $M \geq n+3$ and that no triple of the fixed points is collinear. Then any optimal solution to the generalized minisum hypersphere problem is a hypersphere $S \in \mathscr{G}$.*

Proof. Follows from Theorems 2.9 and 2.13.   □

2.5 Incidence Properties

We now consider incidence properties for the generalized minisum hypersphere problem; that is, the feasible set $\overline{\mathscr{G}}$ consists of all hyperplanes and -spheres in $\mathbb{R}^n$.

Lemma 2.15. *For any $n \geq 2$ there exists at least one optimal solution to the generalized minisum hypersphere problem on $\mathbb{R}^n$ that intersects at least one fixed point.*

Proof. If a hyperplane H is an optimal solution it is known from results for the Euclidean minisum hyperplane problem that H is the affine hull of n affinely independent fixed points (see, e.g., [Sch99]). Therefore we only consider the case where a minisum hypersphere with radius $r < \infty$ exists.

Let $S(X, r) \in \mathscr{G}$ be a minisum hypersphere. Keeping X fixed we obtain a function in r, namely,

$$h(r) = f(S(X, r)) = \sum_{m=1}^{M} \omega_m \big| \|X - A_m\| - r \big|.$$

This is equivalent to a one-dimensional median problem; hence there exists some $m_0 \in \{1, \ldots, M\}$ such that the minimizer r^* of $h(r)$ satisfies

$$r^* = \|X - A_{m_0}\|.$$

Consequently, the corresponding hypersphere $S(X, r^*)$ is still a minisum hypersphere that intersects A_{m_0}.   □

Lemma 2.15 shows that the optimal radius satisfies the *median property*; that is, the sum of weights inside an optimal hypersphere and the sum of weights outside the hypersphere cannot differ too much.

Corollary 2.16 (Median property). *Let $S \in \overline{\mathscr{G}}$ be an optimal solution to the generalized minisum hypersphere problem. If S is a hypersphere then define J_+, J_0, and J_- according to Notation 1.8. If S is a hyperplane then denote the fixed points on either side of S by J_+ and J_-, respectively. Then we have*

$$\sum_{m:A_m \in J_- \cup J_0} \omega_m \geq \sum_{m:A_m \in J_+} \omega_m \quad \text{and} \quad \sum_{m:A_m \in J_+ \cup J_0} \omega_m \geq \sum_{m:A_m \in J_-} \omega_m,$$

or, equivalently,

$$\left| \sum_{m:A_m \in J_-} \omega_m - \sum_{m:A_m \in J_+} \omega_m \right| \leq \sum_{m:A_m \in J_0} \omega_m.$$

Proof. Due to Theorem 2.13 an optimal solution to the generalized minisum hypersphere problem is either a hypersphere or a -plane. In the former case the result is a consequence of the proof of Lemma 2.15. In the latter case the result is known, see [Sch99]. $\qquad\square$

We give a dominance criterion for the minisum hypersphere problem which is analogous to the majority theorem of Witzgall [Wit64] for location of a point. For this purpose we define

$$\text{Test}_m := 2\omega_m - \sum_{m=1}^{M} \omega_m, \quad 1 \leq m \leq M.$$

Lemma 2.17. *Suppose a minisum hypersphere $S(X,r) \in \mathscr{G}$ exists. If there exists a fixed point $A_m \in \mathscr{D}$ such that $\text{Test}_m > 0$, then $S(X,r)$ intersects the fixed point A_m, i.e., $A_m \in S(X,r)$. If $\text{Test}_m = 0$, then there exists at least one optimal solution that intersects A_m.*

Proof. Without loss of generality assume $\text{Test}_1 > 0$. Let $S = S(X,r)$ be a minisum hypersphere and assume $A_1 \notin S$. We show that $A_1 \notin S$ is a contradiction to the optimality of S. To this end we construct a hypersphere $S' = S(X',r)$ such that $f(S') < f(S)$.

Let $Y_m \in S$ such that $\|A_i - Y_i\| = d_m(S)$, $1 \leq m \leq M$. Furthermore, let $X' = X + A_1 - Y_1$ and $S' = S(X',r)$. Then, we have

$$\|Y_m + A_1 - Y_1 - X'\| = \|Y_m - X\| = r$$

for all $m = 1, \ldots, M$; that is, $Y_m + A_1 - Y_1 \in S'$. In particular we obtain $d_1(S') = 0$ and finally

$$
\begin{aligned}
f(S') &= \sum_{m=2}^{M} \omega_m d_m(S') \\
&\leq \sum_{m=2}^{M} \omega_m \|Y_m + A_1 - Y_1 - A_m\| \\
&\leq \sum_{m=2}^{M} \omega_m \|Y_m - A_m\| + \sum_{m=2}^{M} \omega_m \|A_1 - Y_1\| \\
&= \sum_{m=2}^{M} \omega_m d_m(S) + \sum_{m=2}^{M} \omega_m d_1(S) \\
&= f(S) - d_1(S) \left(\omega_1 - \sum_{m=2}^{M} \omega_m \right) \\
&= f(S) - d_1(S) \cdot \text{Test}_1,
\end{aligned}
$$

i.e., $f(S') < f(S)$ if $\text{Test}_1 > 0$. If $\text{Test}_1 = 0$, then $f(S') \leq f(S)$, so that the optimality of S implies that S' is also optimal.   □

A stronger incidence property for the planar case is shown by Brimberg et al. [BJS09b]: *All minisum circles intersect at least two fixed points.* This result is extended to arbitrary dimensions $n \geq 2$ by Nievergelt [Nie10].

Theorem 2.18 ([Nie10]). *All minisum hyperspheres intersect at least two fixed points.*

The proof of Theorem 2.18 is based on the fact that the Laplacian of the objective function $f(S(X,r)$ is strictly negative at the center X if $S(X,r)$ intersects at most one fixed point. The planar version of Theorem 2.18 was stated earlier by Brimberg, Juel, and Schöbel [BJS09b]. Also their proof is based on a strictly negative Laplacian.

For $n = 2$, $M = 4$, and equal weights, a stronger version of Theorem 2.18 applies.

Theorem 2.19 ([Nie10]). *Let $n = 2$, $\mathscr{D} = \{A_1, A_2, A_3, A_4\} \subseteq \mathbb{R}^2$, and assume equal weights $\omega_1 = \ldots = \omega_4$. Then at least one generalized minisum circle intersects three fixed points.*

Like Theorem 2.18, this result is also shown by analysis of second-order derivatives of the objective function f. Note that the result implies that a straight line cannot be an optimal solution, provided that no triple of the fixed points is collinear.

Remark 2.20. Examples in [Sch07] and [Nie10] show that Theorem 2.19 does not extend to instances with more than four fixed points. In particular, the example of Scholz [Sch07] shows that Theorem 2.19 is also not true for more than four fixed points if we additionally suppose that no triple of the fixed points is collinear.

Since the example of Scholz is only available in German we state it here.

Example 2.21 ([Sch07]). Let $n = 2$, and let six fixed points be given such that

$$A_1 = (0,1), \qquad A_2 = (0,-1), \qquad A_3 = (0.8,0),$$
$$A_4 = (-0.8,0), \qquad A_5 = (0.1,0.9), \qquad A_6 = (-0.1,-0.9).$$

Furthermore, let $\omega := \omega_1 = \cdots = \omega_6$ and let C_{ijk} denote the circle intersecting A_i, A_j, and A_k. Elementary calculations show $f(C_{ijk}) > 0.4 \cdot \omega$ for any triple $1 \leq i < j < k \leq 6$. But the circle $C' = C(X', r')$, where $X' = (0,0)$ and $r' = \|A_5\| = \|A_6\|$, has objective value $f(C') = 0.4 \cdot \omega$. Hence, there does not exist a minisum circle intersecting three fixed points. The example is illustrated in Fig. 2.2.

The following example shows that the assumption of equal weights is necessary in Theorem 2.19.

Example 2.22. Let $n = 2$, and let four fixed points be given such that

$$A_1 = (0,6), \qquad A_2 = (0,-6), \qquad A_3 = (-5,0), \qquad A_4 = (5,0).$$

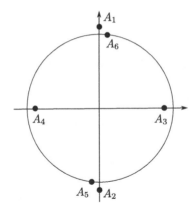

Fig. 2.2 Illustration of Example 2.21

Let $\omega_1 = \omega_2 = 100$, $\omega_3 = \omega_4 = 1$, and let C_{ijk} denote the circle intersecting A_i, A_j, and A_k. Elementary calculations show

$$f(C_{123}) = f(C_{124}) = 2.2 \quad \text{and} \quad f(C_{134}) = f(C_{234}) \approx 183.37.$$

Since the circle $C(X,r)$ with center $X = (0,0)$ and radius $r = 6$ has objective value $f(C(X,r)) = 2$, none of the circles C_{ijk} is a minisum circle.

We close this section with a counterexample disproving the following conjecture.

Conjecture 2.23. The center of a minisum circle is always the intersection point between two bisectors $B(A_i,A_j)$, $B(A_k,A_l)$ of four (not necessary pairwise distinct) fixed points A_i, A_j, A_k, A_l.

Example 2.24. Let $n = 2$, and let four fixed points be given such that

$$A_1 = (0.5,1), \qquad A_2 = (0.5,0), \qquad A_3 = (0.1,0.5), \qquad A_4 = (0.9,0.5).$$

Furthermore, let $\omega_1 = \omega_2 = 100$, $\omega_3 = 1.1$, and $\omega_4 = 1$.

Consider Table 2.1. In the first column the intersection point X between the bisectors B_{ij} and B_{kl} is reported. In the second column the optimal radius r^* corresponding to the center point X is stated which can be easily computed by applying Theorem 2.18. The third column contains the objective value $f(C(X,r^*))$ corresponding to the circle with center X and radius r^*. As it may be obtained from Table 2.1 the best circle has objective value 0.21, its center is at the origin, and it intersects A_1 and A_2. But this circle is not a minisum circle. Indeed, let

$$\overline{X} = \begin{pmatrix} \frac{2+\sqrt{1760}}{2\sqrt{1760}} \\ 0.5 \end{pmatrix} \approx \begin{pmatrix} 0.5238 \\ 0.5 \end{pmatrix}, \qquad \overline{r} = \left(\frac{1}{1760} + \frac{1}{4} \right)^{\frac{1}{2}} \approx 0.5233.$$

Then the circle with center $\overline{X}$ and radius $\overline{r}$ has objective value

$$f(C(\overline{X},\overline{r})) = \frac{11}{\sqrt{110}} - \frac{21}{25} \approx 0.2088,$$

i.e., $C(\overline{X},\overline{r})$ is superior to any circle $C(X,r)$ reported in Table 2.1. $C(\overline{X},\overline{r})$ and the four fixed points A_1, A_2, A_3, and A_4 are depicted in Fig. 2.3. It can be shown that $C(\overline{X},\overline{r})$ is a minisum circle.

Note that Example 2.24 uses different weights for the fixed points. It remains an open question whether or not Conjecture 2.23 is true in case of equal weights. Brimberg et al. prove in [BJS09b] that for equal weights, an unbounded number of fixed points $(M \to \infty)$, and further mild conditions any minisum circle intersects three fixed points. Nievergelt shows in [Nie10] that for equal weights a minisum circle that intersects exactly two fixed points has to satisfy strict conditions. In order

Table 2.1 Objective values of circles intersecting at least two fixed points

X	r^*	$f(C(X,r^*))$
$B_{12} \cap B_{13} = (0.6125, 0.5)$	$\|A_1 - X\| = 0.5125$	0.225
$B_{12} \cap B_{23} = (0.6125, 0.5)$	$\|A_1 - X\| = 0.5125$	0.225
$B_{12} \cap B_{23} = (0.6125, 0.5)$	$\|A_1 - X\| = 0.5125$	0.225
$B_{12} \cap B_{24} = (0.3875, 0.5)$	$\|A_1 - X\| = 0.5125$	0.2475
$B_{12} \cap B_{14} = (0.3875, 0.5)$	$\|A_1 - X\| = 0.5125$	0.2475
$B_{14} \cap B_{24} = (0.3875, 0.5)$	$\|A_1 - X\| = 0.5125$	0.2475
$B_{13} \cap B_{14} = (0.5, 0.59)$	$\|A_1 - X\| = 0.41$	18
$B_{13} \cap B_{34} = (0.5, 0.59)$	$\|A_1 - X\| = 0.41$	18
$B_{14} \cap B_{34} = (0.5, 0.59)$	$\|A_1 - X\| = 0.41$	18
$B_{23} \cap B_{24} = (0.5, 0.41)$	$\|A_2 - X\| = 0.41$	18
$B_{23} \cap B_{34} = (0.5, 0.41)$	$\|A_2 - X\| = 0.41$	18
$B_{24} \cap B_{34} = (0.5, 0.41)$	$\|A_2 - X\| = 0.41$	18
$B_{12} \cap B_{34} = (0, 0)$	$\|A_1 - X\| = 0.5$	0.21
$B_{13} \cap B_{24} = \emptyset$		
$B_{14} \cap B_{23} = \emptyset$		

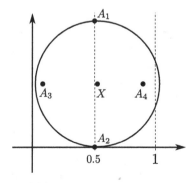

Fig. 2.3 Illustration of Example 2.24

to state these conditions suppose that $C(X,r)$ is a minisum circle that intersects the fixed points $A_1 = (0,y)$ and $A_2 = (0,-y)$. Then we have $X = (t,0)$ for some $t \in \mathbb{R}$ and t must satisfy

$$\sum_{m:A_m \in J_+} \frac{A_{m1} - t}{\sqrt{(A_{m1} - t)^2 + A_{m2}^2}} = \sum_{m:A_m \in J_-} \frac{A_{m1} - t}{\sqrt{(A_{m1} - t)^2 + A_{m2}^2}}$$

and

$$\sum_{m:A_m \in J_+} \frac{(A_{m1} - t)^2}{(A_{m1} - t)^2 + A_{m2}^2} \neq \sum_{m:A_m \in J_-} \frac{(A_{m1} - t)^2}{(A_{m1} - t)^2 + A_{m2}^2}$$

where $A_m = (A_{m1}, A_{m2})$ for all $1 \leq m \leq M$.

2.6 Solution Approaches for the Planar Case

Theorem 2.18 gives rise to a solution approach for the Euclidean minisum circle problem. According to this result the center of any minisum circle must lie on the bisector of two fixed points. Thus, the search for an optimal solution may be reduced to a series of one-dimensional searches along all bisectors defined by the fixed points $\mathscr{D}$. This search can be done by performing two steps. Firstly, for all circles passing through three fixed points and satisfying the median property (cf. Corollary 2.16) compute the corresponding objective value. Subsequently, search for local minima of f on each open segment $I \subseteq B_{ij}$, $1 \leq i < j \leq M$, that does not intersect another bisector. Following this approach all minisum circles may be reported in $\mathscr{O}(M^4)$ operations (cf. [BJS09b]).

If the fixed points $\mathscr{D}$ have equal weights, are in general position, and have the property that at most three distinct fixed points lie on a circle, then it is possible to exclude segments $I \subseteq B_{ij}$, $1 \leq i < j \leq M$, from the search, see [Nie10]. However, the worst case complexity of the search remains unchanged. If the fixed points do not satisfy the conditions of Theorem 2.9 then also all straight lines through two distinct fixed points have to be evaluated. This can be done in $\mathscr{O}(M^2)$ operations, see Korneenko and Martini [KM90, KM93, MS98].

A natural heuristic approach for the Euclidean minisum circle problem consists of evaluating all circles passing through three distinct fixed points. In [BJS09b] a numerical study is presented where 500 location problems with equal weights are investigated. In 498 cases the heuristic was able to find an optimal solution. The relative error of the heuristic in the two remaining cases is smaller than 0.01%. For weighted sets of fixed points no results are reported in [BJS09b].

Whereas the exact method proposed by Brimberg et al. [BJS09b] and Nievergelt [Nie10] is not appropriate for large instances, it turned out that the *big cube small cube method* also solves large instances very efficiently. The big cube small cube method is a geometric branch and bound scheme where a (bounded) solution space is decomposed into (hyper)cubes. For each cube, lower and upper bounds are computed and used in order to discard cubes. If a cube cannot be discarded, then it is split into smaller cubes. The method ends with an ε-approximation of an optimal solution where $\varepsilon > 0$ can be chosen arbitrarily, see [SS10] for further details. Schöbel and Scholz [SS10] apply the big cube small cube method to a restricted version of the Euclidean minisum circle problem where center X and radius r have to lie in a compact box $B \subseteq \mathbb{R}^3$. They study weighted instances with up to 10,000 fixed points and solve each in less than 11 s with an accuracy of $\varepsilon = 10^{-10}$. The big cube small cube method can also be applied to minisum hypersphere problems, i.e., for higher dimensions $n \geq 3$. But in this case the number of hypercubes increases exponentially in n and therefore the performance of the method decreases. The Euclidean minisum circle location problem is also studied by Blanquero et al. [BCH09]. They use a representation of the objective function as difference of convex functions and DC programming (see, e.g., [AT05]) in order to approximate minisum circles.

2.7 Concluding Remarks

From a practical point of view the planar version of the Euclidean hypersphere problem is solved. In most cases the conditions of Theorem 2.9 are met by the fixed points and an upper bound for the radius can be obtained. Thus, all (restricted) minisum circles may be approximated with arbitrary accuracy using the big cube small cube method. However, from a mathematical point of view there remain open questions. As mentioned above it is an open question whether or not Conjecture 2.23 is true for fixed points with equal weights. Another open point is an estimation of the difference between the objective value of a minisum circle and the objective value of the best circle passing through three fixed points. As Nievergelt already noted in [Nie10], an efficient algorithm analog to Korneenko and Martini's *Anchored Median Hyperplane Algorithm* [KM90, KM93, MS98] for the task of computing the best circle through three fixed points is also of interest. Further possible lines of research may include the *multicircle problem* where more than one circle is sought and *obnoxious facility location problems* where some fixed points have negative weights.

For arbitrary dimensions $n \geq 3$ an efficient algorithm for the Euclidean minisum hypersphere problem is not available. The geometric descriptions known for minisum hyperspheres for $n \geq 3$ are not sufficient in order to give rise to a solution method. Thus, further analysis of minisum hyperspheres is required. Alternatively, general approaches of global optimization may be used in order to approximate minisum hyperspheres in a practical setting.

An interesting special case of the Euclidean minisum circle problem is the case where the radius is fixed. In this case the only decision variable is the center $X \in \mathbb{R}^2$. Brimberg et al. [BJS09b] studied this problem. In contrast to the unrestricted problem where any minisum circle intersects at least two fixed points they showed that an optimal circle for the problem with fixed radius need not intersect any of the fixed points. Furthermore, they showed that for a sufficiently small radius $\bar{r}$ there exists a strong relation between the Euclidean minisum circle problem with fixed radius $\bar{r}$ and the Weber problem: Let $X \in \mathbb{R}^2$ be an optimal solution to the Weber problem with weighted fixed points $\mathcal{D}$. Then $C(X,\bar{r})$ is an optimal solution to the minisum circle problem with radius $\bar{r}$, provided that $d(A,X) \geq r$ for all $A \in \mathcal{D}$. Analog results for higher dimensions $n \geq 3$ seem possible, but require further analysis.

Another problem related to the Euclidean minisum hypersphere problem is the location of an $(n-1)$-dimensional hypersphere on an n-dimensional hypersphere. For the case $n = 3$ this problem is studied in [BJS07]. It is shown that at least one optimal circle for this problem exists passing through one fixed point. The stronger incidence property that any optimal circle passes through at least two fixed points is an open question. For the special case of locating great circles on a sphere this incidence property applies (cf. [BJS09b]).

Also of interest is the problem of locating a circle in an n-dimensional space.

Chapter 3
Minisum Hyperspheres in Normed Spaces

3.1 Basic Assumptions

In this section we analyze properties of the minisum hypersphere problem in a normed plane, i.e., we consider the following setting:

Dimensions: $n \geq 2, M \geq 2$
Norm: Arbitrary norm $\|\cdot\|$ in $\mathbb{R}^n$
Objects: Hyperspheres defined with respect to $\|\cdot\|$
Metric: Induced by $\|\cdot\|$; $d(X,Y) := \|Y - X\|$
Distance: $d(S,A_m) = \min\{d(Y,A_m) : Y \in S\}, 1 \leq m \leq M$
Objective: $f(S) = \sum_{m=1}^{M} \omega_m d(S,A_m) \to \min$

All results presented in this chapter assume that the above setting is satisfied. In addition we assume without exception that at least three fixed points are given, i.e., $M \geq 3$. For $M \leq 2$ the minisum hypersphere problem is trivial.

There are a lot of results which are valid for the Euclidean norm but do not extend to arbitrary norms. For instance, arbitrary norms do not have perpendicular bisectors. Even more, a bisector of an arbitrary norm may contain two-dimensional regions as we have seen in Section 1.3. In this chapter we discuss which results for the Euclidean case are still valid for arbitrary norms and which results do not hold true for arbitrary norms. The two-dimensional results presented in this chapter are based on some joint work with Brimberg, Juel, Schöbel, and the author, see [BJKS09].

3.2 Distance

As for the Euclidean case there exists a simple formula for the point–hypersphere distance.

M.-C. Körner, *Minisum Hyperspheres*, Springer Optimization and Its Applications 51,
DOI 10.1007/978-1-4419-9807-1_3, © Springer Science+Business Media, LLC 2011

Lemma 3.1. *The distance between a hypersphere $S(X,r)$ and a point A is given by*

$$d(S(X,r),A) = |\,\|X - A\| - r|.$$

Proof. Only properties of a norm were used in the proof of Lemma 2.1. Therefore, the result may be obtained analogously to Lemma 2.1.   □

As a consequence we obtain the following convexity properties.

Corollary 3.2. *Let $A, X \in \mathbb{R}^n$. Then the point–hypersphere distance $d(S(X,r),A)$ is convex and piecewise linear in r.*

Corollary 3.3. *Let $X, A \in \mathbb{R}^n$. Then the distance $d(S(X,r),A)$ is*

(i) concave in (X,r) on the set

$$\mathcal{V} = \{(X,r) \in \mathbb{R}^n \times]0,\infty[:\ \|X - A\| \le r\}$$

  and
(ii) convex in (X,r) on any convex set $U \subseteq (\mathbb{R}^n \times]0,\infty[) \setminus \mathcal{V}$.

The following symmetry property applies.

Corollary 3.4. *Let $A \in \mathbb{R}^n$ and $S(X,r) \in \mathcal{G}$. Then we have $S(A,r) \in \mathcal{G}$ and*

$$d(S(X,r),A) = d(S(A,r),X).$$

Proof. $S(A,r) \in \mathcal{G}$ is clear. Therefore $d(S(A,r),X)$ is well-defined and the result follows from Lemma 3.1.   □

Note that in Section 2.2 exactly the same results were obtained for the Euclidean case.

3.3 Degenerated Solutions

Independent of the norm $\|\cdot\|$, a hypersphere degenerated to a point (radius $r = 0$) cannot be a minisum hypersphere.

Lemma 3.5. *Any hypersphere $S(X,r)$ with radius $r = 0$ cannot be a minisum hypersphere.*

Proof. The result may be obtained analogously to Lemma 2.5 by substituting the Euclidean norm in the proof for an arbitrary norm $\|\cdot\|$.   □

For the Euclidean case, we have shown in Lemma 2.8 that another extreme is possible: a minisum hypersphere can be degenerated to a hyperplane. More precisely, for any dimension $n \ge 2$ there exist a set of fixed points and a hyperplane H such that the objective value of H is superior to the objective value of any Euclidean hypersphere. Thus, in general the set of all Euclidean hyperspheres does not contain a minisum hypersphere. The following result shows that for any norm $\|\cdot\|$ a hyperplane can exist which is superior to all hyperspheres (defined with respect to norm $\|\cdot\|$). For the class of strictly convex norms and the class of smooth norms this implies that it is possible that no minisum hypersphere exists.

Lemma 3.6. *Let $\|\cdot\|$ be any norm in $\mathbb{R}^n$. Then for each $n \geq 2$ there exist a finite set $\mathscr{D} \subseteq \mathbb{R}^n$ and a hyperplane H such that the objective value of H is superior to the objective value of any hypersphere. Furthermore, for smooth or strictly convex norms there exist finite sets $\mathscr{D} \subseteq \mathbb{R}^n$ such that no hypersphere is an optimal solution to the corresponding minisum hypersphere problem with fixed points $\mathscr{D}$.*

Proof. Let $H \subseteq \mathbb{R}^n$ be a hyperplane and choose a finite set $\mathscr{D}$ such that $\mathscr{D} \subseteq H$ and $|\mathscr{D}| > 2$. Then the distance between H and each fixed point $A_m \in \mathscr{D}$ is zero; hence, $f(H) = 0$. If $\|\cdot\|$ is smooth or strictly convex, then there does not exist a hypersphere $S(X,r)$ such that $\mathscr{D} \subseteq S(X,r)$; hence, $f(S(X,r)) > 0$. If $\|\cdot\|$ is a polyhedral norm, then we cannot choose H arbitrarily. Instead, we choose H such that H is not parallel to a flat spot of the unit ball of $\|\cdot\|$. Then we may also conclude $f(S(X,r)) > 0$ for all $S(X,r) \in \mathscr{G}$.

If $\|\cdot\|$ is smooth or strictly convex, then we can choose H such that a sequence of hyperspheres exists which converges against H. Therefore, the second part of the assertion may be obtained analogously to Lemma 2.8. $\qquad\square$

The idea of Lemma 3.6 is to choose the set of fixed points $\mathscr{D}$ in such a way that $\mathscr{D}$ is contained in a hyperplane $H \subseteq \mathbb{R}^n$ but not in any hypersphere $S \in \mathscr{G}$. However, the following example shows that for smooth norms degenerated solutions are possible even if the fixed points are in general position.

Example 3.7. Assume $n = 2$, let $\|\cdot\|$ be the norm we obtain by rounding the vertices of a paraxial rectangle, and consider the six fixed points outlined in Fig. 3.1 with weights $\omega_1 = 10$, $\omega_2 = 10$, and $\omega_3 = \cdots = \omega_6 = 1$. Due to the dominance of the weights of the fixed points A_1 and A_2 the line $\ell_{12} = \langle A_1, A_2 \rangle$ is an optimal solution to the corresponding line location problem, see [Sch99]. We have $f(\ell_{12}) = 8$, since $\min_{Y \in \ell_{12}} \|A_m - Y\| = 2$ for all $3 \leq m \leq 6$ (see the unit circle $C(X_0, 1)$ depicted in Fig. 3.1). Analogously, we may read from Fig. 3.1 that $f(C(X_1, 6)) = 8$.

We show that no circle $C \in \mathscr{G}$ exists which is superior to $C(X_1, 6)$. To this end consider the circles $C_1 = C(A_1, 0.8)$ and $C_2 = C(A_2, 0.8)$ depicted in Fig. 3.1. Since $f(C(X_1, 6)) = 8$ any circle C superior to $C(X_1, 6)$ has to intersect the circles C_1 and C_2; otherwise we have $f(C) \geq \omega_m d_m(C) > 8$ for $m = 1, 2$. But any circle intersecting C_1 and C_2 has objective value

$$f(C) \geq \sum_{m=3}^{M} \omega_m d_m(C) \geq 8.$$

Hence, we may conclude that no circle $C \in \mathscr{G}$ exists which is superior to $C(X_1, 6)$. In particular, the straight line ℓ_{12} is a degenerated solution which has the same objective value as the best circle $C \in \mathscr{G}$.

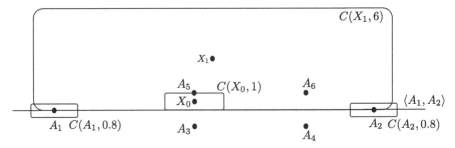

Fig. 3.1 Circle location problem with smooth but not strictly convex norm

Using the notion of the proof of Lemma 2.8 we can derive a weaker result for polyhedral norms and smooth but not strictly convex norms.

Lemma 3.8. *Let* $\| \cdot \|$ *be a polyhedral norm or a smooth but not strictly convex norm. Then for each $n \geq 2$ there exist a finite set $\mathscr{D} \subseteq \mathbb{R}^n$ and a hyperplane H such that the objective value of H is equal to the objective value of a minisum hypersphere $S(X,r) \in \mathscr{G}$.*

Proof. Let $\mathscr{D} \subseteq \mathbb{R}^n$ be a finite set of fixed points contained in a hyperplane $H \subseteq \mathbb{R}^n$ which is parallel to a flat spot of the unit ball of the norm $\| \cdot \|$. On the one hand, the distance between H and the fixed points vanishes and therefore we have $f(H) = 0$. On the other hand, due to the construction of $\mathscr{D}$ there has to be a hypersphere $S(X,r) \in \mathscr{G}$ such that $\mathscr{D} \subseteq S(X,r)$. We conclude $f(S(X,r)) = f(H)$; in particular, we may also conclude that $S(X,r)$ is a minisum hypersphere.   □

For the class of elliptic norms mild conditions exist guaranteeing that any hyperplane is inferior to a minisum hypersphere.

Theorem 3.9. *Let $\mathscr{D} \subseteq \mathbb{R}^n$ be in general position. Suppose $M \geq n + 3$ and that no triple of the fixed points is collinear. If $\| \cdot \|$ is an elliptic norm, then there does not exist a hyperplane H such that*

$$f(H) \leq f(S) \text{ for all } S \in \mathscr{G}.$$

Proof. Let the hyperplane $H \subseteq \mathbb{R}^n$ be an optimal solution to the minisum hyperplane location problem. Then H is the affine hull of n affinely independent fixed points, see Theorem 7.4 in [Sch99]. Thus, H intersects at least n fixed points. Due to the assumptions of the theorem, H intersects exactly n fixed points, say $A_1, \ldots, A_n$. We now construct two hyperspheres $S_a = S(X_a, r)$ and $S_b = S(X_b, r)$ having equal radius r such that either $f(S_a) < f(H)$ or $f(S_b) < f(H)$. This proves the result since $f(H) \leq f(H')$ for all hyperplanes $H' \subseteq \mathbb{R}^n$.

In order to construct these hyperspheres denote by J_1 and J_2 the fixed points in either half-space defined by H; that is,

$$\mathscr{D} = \{A_1, \ldots, A_n\} \cup J_1 \cup J_2.$$

Let P_m be the projection of A_m on H, $m = n+1, n+2, \ldots, M$, such that

$$d(A_m, H) = \min_{Y \in H} \|A_m - Y\| = \|A_m - P_m\|.$$

Because an elliptic norm is strictly convex, it follows that P_m is uniquely determined for each given A_m. Using the fact that an elliptic norm is also smooth we may apply Theorem 1.25 and obtain that there exists a $(n-1)$-dimensional hypersphere $S^{n-1} \subseteq H$ such that $A_m \in S^{n-1}$ for all $1 \le m \le n$. Hence, we may construct two hyperspheres $S_a = S(X_a, r)$, $S_b = S(X_b, r)$ of the same radius, each intersecting $A_1, A_2, \ldots, A_n$ such that X_a, X_b are located in distinct half-spaces separated by H and

$$\|X_a - A_i\| = \|X_a - A_j\| \quad \text{for all } 1 \le i < j \le n,$$

$$\|X_a - A_i\| = \|X_b - A_i\| \quad \text{for all } 1 \le i \le n.$$

Applying Theorem 1.25 on the $n+1$ points $A_1, \ldots, A_n, Y$ where $Y \notin H$ implies that we can make the radius r sufficient large so that $J_-(S_a) = J_+(S_b) = J_1$ and $J_+(S_a) = J_-(S_b) = J_2$. Thus, we may obtain a constellation as depicted in Fig. 3.2 for the planar case.

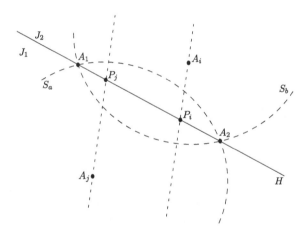

Fig. 3.2 Two-dimensional illustration of the hyperspheres S_a and S_b constructed in the proof of Theorem 3.9

Note that $A_1, \ldots, A_n$ lie on the bisector $B(X_a, X_b)$ of X_a and X_b. Since the bisector of an elliptic norm is a hyperplane, see Theorem 1.52, it follows that $B(X_a, X_b) = H$. This means $\|P - X_a\| = \|P - X_b\|$ for all $P \in H$.

We now investigate the distance from a point $A_j \in J_1$ to the hyperspheres S_a and S_b and compare it with the distance of A_j to the hyperplane H. Due to the triangle inequality we have $\|A_j - X_b\| \le \|A_j - P_j\| + \|P_j - X_b\|$ and $\|X_a - P_j\| \le \|X_a - A_j\| + \|A_j - P_j\|$. Thus, we obtain

$$
\begin{aligned}
d_j(S_a) + d_j(S_b) &= (r - \|A_j - X_a\|) + (\|A_j - X_b\| - r) \\
&= \|A_j - X_b\| - \|A_j - X_a\| \\
&\le \|A_j - P_j\| + \|P_j - X_b\| + \|A_j - P_j\| - \|P_j - X_a\| \\
&= 2\|A_j - P_j\| = 2d(H, A_j).
\end{aligned}
$$

Analogously, we obtain

$$
d_j(S_a) + d_j(S_b) \le 2\|A_j - P_j\| = 2d(H, A_j)
$$

for $A_j \in J_2$. Since the elliptic norm $\|\cdot\|$ is strictly convex, the assumptions on collinearity and number of fixed points ($M \ge n+3$) imply that at least one of these inequalities is strict. Thus, we finally obtain

$$
\begin{aligned}
&f(S_a) + f(S_b) \\
&= \sum_{m \in J_1} \omega_m (d_m(S_a) + d_m(S_b)) + \sum_{m \in J_2} \omega_m (d_m(S_a) + d_m(S_b)) \\
&< \sum_{m=1}^{M} \omega_m 2d(H, A_m) \\
&= 2f(H);
\end{aligned}
$$

that is, $f(S_a) < f(H)$ or $f(S_b) < f(H)$.   □

3.4 Existence of Minisum Hyperspheres

If $\|\cdot\|$ is a polyhedral norm, then a hyperplane may have the same objective value as a minisum hypersphere, see Lemma 3.8. Nevertheless, for polyhedral norms we can show that $\mathscr{G}$ always contains a minisum hypersphere.

Theorem 3.10. *Let $\|\cdot\|$ be a polyhedral norm. Then $\mathscr{G}$ contains a minisum hypersphere.*

Proof. Let B denote the unit ball of the polyhedral norm $\|\cdot\|$. Define

$$
\Upsilon_B := \{(\mathscr{F}_1, \ldots, \mathscr{F}_M) : \mathscr{F}_m \text{ facet of } B, \; 1 \le m \le M\}.
$$

Furthermore, for any $(\mathscr{F}_1, \ldots, \mathscr{F}_M) \in \Upsilon_B$ define

$$
\mathscr{C}(\mathscr{F}_1, \ldots, \mathscr{F}_M) := \bigcap_{m=1}^{M} (\{A_m\} + \Gamma_{\mathscr{F}_m})
$$

where $\Gamma_{\mathscr{F}_m}$ is a fundamental cone of norm $\|\cdot\|$ (cf. Notation 1.39). In our case of a polyhedral norm $\|\cdot\|$ the sets $\mathscr{C}(\mathscr{F}_1,\ldots,\mathscr{F}_M)$ are what Durier and Michelot call *elementary convex sets* (cf. [DM85]). Note that each set $\mathscr{C}(\mathscr{F}_1,\ldots,\mathscr{F}_M)$ is a (possibly unbounded) convex polyhedron in $\mathbb{R}^n$. Furthermore, we have

$$\bigcup_{(\mathscr{F}_1,\ldots,\mathscr{F}_M)\in \Upsilon_B} \mathscr{C}(\mathscr{F}_1,\ldots,\mathscr{F}_M) = \mathbb{R}^n;$$

that is, the sets $\mathscr{C}(\mathscr{F}_1,\ldots,\mathscr{F}_M)$ induce a finite tessellation of $\mathbb{R}^n$ into convex polyhedrons. In the following we show that the minisum hypersphere problem with polyhedral norm $\|\cdot\|$ has a (local) optimum on each set $\mathscr{C}(\mathscr{F}_1,\ldots,\mathscr{F}_M)\times]0,\infty[$; that is, we show that there exists $(X,r) \in \mathscr{C}(\mathscr{F}_1,\ldots,\mathscr{F}_M)\times]0,\infty[$ such that $S(X,r)$ is an optimal solution to the minisum hypersphere problem with solution set

$$\mathscr{G}_{\mathscr{F}_1,\ldots,\mathscr{F}_M} := \{S(X,r) : X \in C(\mathscr{F}_1,\ldots,\mathscr{F}_M),\ r\in]0,\infty[\} \subseteq \mathscr{G}.$$

It follows that $\mathscr{G}$ has to contain a minisum hypersphere.

Choose an arbitrary set $\mathscr{C}(\mathscr{F}_1,\ldots,\mathscr{F}_M) \subseteq \mathbb{R}^n$. For any fundamental cone $\Gamma_{\mathscr{F}_m}$ let $E_{m1},\ldots,E_{mt_m},\, t_m \geq n$, denote the extreme points of the corresponding facet of B. Consider the following linear program:

$$
\begin{array}{lll}
\text{Minimize} & \sum_{m=1}^{M} \omega_m(z_m^+ + z_m^-) & \\
\text{subject to} & \sum_{g=1}^{t_m} \beta_{mg} - r = z_m^+ - z_m^- & (1 \leq m \leq M) \\
& \sum_{g=1}^{t_m} \beta_{mg}E_{mg} = X - A_m & (1 \leq m \leq M) \\
& X \in \mathscr{C}(\mathscr{F}_1,\ldots,\mathscr{F}_M) & \\
& \beta_{mg} \geq 0 & (1 \leq m \leq M,\, 1 \leq g \leq t_m) \\
& z_m^+, z_m^-, r \geq 0 & (1 \leq m \leq M)
\end{array}
\qquad \text{(LP)}
$$

Since $X \in \mathscr{C}(\mathscr{F}_1,\ldots,\mathscr{F}_M)$ we have $X \in \{A_m\} + \Gamma_{\mathscr{F}_m}$ for all $A_m \in \mathscr{D}$. In particular, $X - A_m \in \Gamma_{\mathscr{F}_m}$ for all $A_m \in \mathscr{D}$. Hence, we obtain from Lemma 1.40 that for any representation of $X - A_m = \sum_{g=1}^{t_m} \beta_{mg}E_{mg}$ in terms of $E_{m1},\ldots,E_{mt_m}$ we have $\|X - A_m\| = \sum_{g=1}^{t_m} \beta_{mg}$. Thus, for any feasible solution to (LP) we have $z_m^+ + z_m^- = d_m(S(X,r)),\, 1 \leq m \leq M$. It follows that (LP) is equivalent to the restricted minisum hypersphere problem with solution set $\mathscr{G}_{\mathscr{F}_1,\ldots,\mathscr{F}_M}$. From $z_m^+, z_m^-, \omega_m \geq 0$ it follows that (LP) is bounded below by 0. Hence, (LP) has an optimal solution, provided that $C(\mathscr{F}_1,\ldots,\mathscr{F}_M)$ is not empty. $\square$

Recall that we have shown in Lemma 3.6 that a result analogous to Theorem 3.10 is not possible for smooth or strictly convex norms; that is, for smooth or strictly convex norms the set $\mathscr{G}$ of all hyperspheres in $\mathbb{R}^n$ does not always contain a minisum hypersphere. For the Euclidean case this problem was solved by defining the set $\overline{\mathscr{G}}$ of *generalized hyperspheres* which consists of all Euclidean hyperspheres and hyperplanes in $\mathbb{R}^n$. It can be shown that $\overline{\mathscr{G}}$ always contains a minimizer for the *generalized* minisum hypersphere problem with Euclidean norm, see Theorem 2.13. The definition of generalized hyperspheres may be extended to arbitrary norms but the approach leading to Theorem 2.13 cannot be applied to smooth or strictly convex norms. For strictly convex norms the set of generalized hyperspheres is not large enough, i.e., there exist instances where neither a hypersphere nor a hyperplane is an optimal solution, as Example 3.11 shows. But also for smooth norms we cannot adapt the Euclidean approach since an algebraic relation between hyperspheres (defined with respect to the norm $\|\cdot\|$) and hyperplanes analog to (2.4) on p. 28 is not known.

Example 3.11. Assume $n = 2$ and let $\|\cdot\|$ be the norm we obtain from the Euclidean unit ball by deleting the strip between the lines

$$\left\langle \begin{pmatrix} 1/\sqrt{2} \\ 1/\sqrt{2} \end{pmatrix}, \begin{pmatrix} -1/\sqrt{2} \\ 1/\sqrt{2} \end{pmatrix} \right\rangle, \qquad \left\langle \begin{pmatrix} 1/\sqrt{2} \\ -1/\sqrt{2} \end{pmatrix}, \begin{pmatrix} -1/\sqrt{2} \\ -1/\sqrt{2} \end{pmatrix} \right\rangle,$$

and then gluing together the upper and lower remaining parts. The outcome of this construction is a strictly convex norm with two vertices and the property that the (limiting) tangents in the vertices have slopes 1 and -1. Now, consider two fixed points A_1, A_2 and a third point A_3 which is the intersection between the straight line ℓ_1 through A_1 with slope 1 and the straight line ℓ_2 through A_2 with slope -1, see Fig. 3.3. Note that no circle $C \in \mathcal{G}$ exists that passes through A_1, A_2, and A_3 (otherwise ℓ_1 and ℓ_2 cannot be tangents to the strict convex circle C). Furthermore, note that the bent line through A_1, A_2, A_3 is the local limit of a sequence of circles $C \in \mathcal{G}$ with increasing radius. Hence, this bent line is a limit of a sequence of circles and it has objective value zero. In particular neither a circle nor a straight line can have the same objective value and therefore neither a circle nor a straight line is a generalized minisum circle.

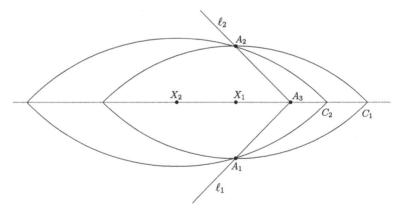

Fig. 3.3 Illustration of Example 3.11

In order to ensure the existence of a minisum hypersphere we choose a pragmatic approach: We do not extend the set $\mathcal{G}$ but we restrict it to a smaller set.

Notation 3.12. Given $\bar{r} > 0$ define

$$\mathcal{G}_{\bar{r}} = \{S(X,r) \in \mathcal{G} : 0 < r \leq \bar{r}, \ S(X,r) \cap \mathrm{conv}(\mathcal{D}) \neq \emptyset\} \subseteq \mathcal{G}.$$

We denote the minisum hypersphere problem with feasible set $\mathcal{G}_{\bar{r}}$ as *restricted minisum hypersphere problem*. An optimal solution to this problem is called *restricted minisum hypersphere*.

The following results justify the restriction to $\mathcal{G}_{\bar{r}}$.

Lemma 3.13. *Let* $\|\cdot\|$ *be an arbitrary norm and suppose that* $\mathcal{G}$ *contains a minisum hypersphere* $S(X,r)$ *with radius* $r < \bar{r}$. *Then we have* $S(X,r) \in \mathcal{G}_{\bar{r}}$.

Proof. Let $S(X,r) \in \mathcal{G}$ be a minisum hypersphere with radius $r < \bar{r}$. If $S(X,r)$ does not intersect the convex hull of the fixed points, then $\mathrm{conv}(\mathcal{D})$ either lies within or outside the hypersphere $S(X,r)$. If $\mathrm{conv}(\mathcal{D})$ lies within $S(X,r)$, then we keep X fixed and decrease r. If $\mathrm{conv}(\mathcal{D})$ lies outside $S(X,r)$, then we increase r. In both cases we obtain a contradiction to the optimality of S. Hence, $S(X,r)$ has to intersect the convex hull of the fixed points and we obtain $S(X,r) \in \mathcal{G}_{\bar{r}}$. $\square$

Corollary 3.14. *Let* $\|\cdot\|$ *be an arbitrary norm. If* $\mathcal{G}$ *contains a minisum hypersphere* $S(X,r)$, *then there exists* $\bar{r} > 0$ *such that* $S(X,r) \in \mathcal{G}_{\bar{r}}$.

Lemma 3.15. *There always exists a restricted minisum hypersphere for the restricted minisum hypersphere problem with arbitrary norm* $\|\cdot\|$.

Proof. Note that

$$\mathcal{U} = \{X \in \mathbb{R}^n : \min_{Y \in \mathrm{conv}(\mathcal{D})} \|X - Y\| \leq \bar{r}\} \times [0, \bar{r}]$$

is closed and bounded and therefore a compact subset of $\mathbb{R}^n \times \mathbb{R}$. Due to the fact that $g(X,r) := f(S(X,r))$ is continuous in X and r on $\mathbb{R}^n \times [0, \infty[$, see Lemma 3.1, we may conclude that $g(X,r)$ attains a minimum on $\mathcal{U}$, say (X^*, r^*). Since

$$\mathcal{G}_{\bar{r}} \subseteq \{S(X,r) : (X,r) \in \mathcal{U}\}$$

it remains to show $S(X^*, r^*) \in \mathcal{G}_{\bar{r}}$. To this end assume the opposite; that is, assume $S(X^*, r^*) \notin \mathcal{G}_{\bar{r}}$. Then we may distinguish two cases:

(i) $S(X^*, r^*) \cap \mathrm{conv}(\mathcal{D}) = \emptyset$ and $\bar{r} > r^* > 0$
(ii) $r^* = 0$

In case (i) we have $S(X^*, r^*) \in \mathcal{G}$ and $r^* < \bar{r}$, hence we may apply Lemma 3.13 and obtain a contradiction. Using Lemma 3.5 we also obtain a contradiction for case (ii); a minisum hypersphere cannot have radius $r = 0$. $\square$

In summary, if $\mathcal{G}$ contains a minisum hypersphere then it is always possible to choose $\bar{r}$ in such a way that also $\mathcal{G}_{\bar{r}}$ contains this minisum hypersphere. If $\mathcal{G}$ does not

contain a minisum hypersphere then $\mathcal{G}_{\bar{r}}$ even so contains an optimal solution which
is a local minimum of the minisum hypersphere problem. Beyond that, numerical
results show that for a set of fixed points $\mathcal{D}$ in general position it is unlikely that $\mathcal{G}$
does not contain a minisum hypersphere. Thus, the restriction of $\mathcal{G}$ to $\mathcal{G}_{\bar{r}}$ may be
less important in a practical setting.

3.5 Incidence Properties

In this section we study incidence properties for minisum hyperspheres in $\mathcal{G}$. As
mentioned in the previous section, the existence of a minisum hypersphere in $\mathcal{G}$ is
ensured, provided that $\|\cdot\|$ is a polyhedral norm.

Using Lemma 3.1 we obtain the following result.

Corollary 3.16. *Let* $\|\cdot\|$ *be a norm in* $\mathbb{R}^n$ *and suppose that* $\mathcal{G}$ *contains a minisum
hypersphere. Then there exists at least one minisum hypersphere intersecting at least
one fixed point.*

Proof. Let $S(X,r) \in \mathcal{G}$ be a minisum hypersphere. Keeping X fixed we obtain a
function in r, namely,

$$h(r) = f(S(X,r)) = \sum_{m=1}^{M} \omega_m \big| \|X - A_m\| - r \big|.$$

This is equivalent to a one-dimensional median problem; hence, there exists some
$m_0 \in \{1,\ldots,M\}$ such that the minimizer r^* of $h(r)$ satisfies

$$r^* = \|X - A_{m_0}\|.$$

Consequently, the corresponding hypersphere $S(X, r^*)$ intersects A_{m_0}. $\square$

The following example shows that Corollary 3.16 does not hold for the restricted
minisum hypersphere problem.

Example 3.17. Assume $n = 2$, $\bar{r} = 1$, and consider the following example using the
Manhattan norm, with $M = 3$ fixed points and weights given as follows:

m	A_m	ω_m
1	$(2,0)$	1
2	$(0,1)$	1
3	$(0,-1)$	1

The unique restricted minisum circle is the circle $C(O,1)$ with radius 1 having its
center at the origin O. $C(O,1)$ does not contain any fixed point; that is, $C(O,1)$ does
not have the median property. An illustration of this example is depicted in Fig. 3.4.

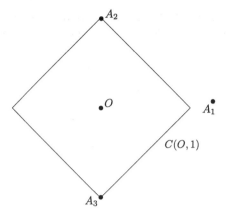

Fig. 3.4 Illustration of Example 3.17

Corollary 3.16 shows that the radius of a minisum hypersphere satisfies the *median property*, namely, that the sum of weights inside a minisum hypersphere and the sum of weights outside a minisum hypersphere cannot differ too much.

Corollary 3.18 (Median property). *Let* $\|\cdot\|$ *be a norm in* $\mathbb{R}^n$ *and suppose that* $\mathscr{G}$ *contains a minisum hypersphere* $S(X,r) \in \mathscr{G}$ *and define* J_+, J_-, J_0 *according to Notation 1.8. Then we have*

$$\sum_{m:A_m \in J_- \cup J_0} \omega_m \geq \sum_{m:A_m \in J_+} \omega_m \quad and \quad \sum_{m:A_m \in J_+ \cup J_0} \omega_m \geq \sum_{m:A_m \in J_-} \omega_m,$$

or, equivalently,

$$\left| \sum_{m:A_m \in J_-} \omega_m - \sum_{m:A_m \in J_+} \omega_m \right| \leq \sum_{m \in J_0} \omega_m.$$

Remark 3.19. Note that Lemma 1.5 (see Section 1.2.2) is a consequence of Corollary 3.18.

The median property implies the following generalization of Lemma 3.13.

Corollary 3.20. *For any norm* $\|\cdot\|$, *if a hypersphere* $S \in \mathscr{G}$ *does not intersect the convex hull of the fixed points* $\mathscr{D}$, *then* S *is not a minisum hypersphere.*

Proof. Let $S(X,r) \in \mathscr{G}$ and assume that $S(X,r) \cap \text{conv}(\mathscr{D}) = \emptyset$. Then $S(X,r)$ does not satisfy the median property. Hence, it cannot be a minisum hypersphere, see Corollary 3.18. □

In the Euclidean case all minisum hyperspheres intersect at least two fixed points (see Theorem 2.18). This incidence property is not true for the case of an arbitrary norm as the following counterexample demonstrates.

Example 3.21. Assume $n = 2$ and consider the following small illustrative example using the Manhattan norm, with $M = 4$ fixed points and weights given as follows:

m	A_m	ω_m
1	$(-0.5, -0.5)$	10
2	$(-0.5, 0.25)$	0.5
3	$(0, -1.1)$	1
4	$(-1.1, 0)$	1

The optimal solution after enumeration (using results described in Section 3.6) is found to be $C_1 = C((0,0),1)$, or any circle $C((t,0),t+1)$ for $t \geq 0$. Each of these circles only intersects the point A_1 and none of the other fixed points, see Fig. 3.5.

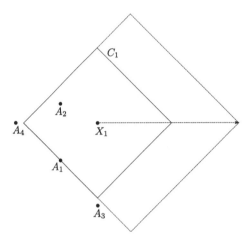

Fig. 3.5 Illustration of Example 3.21. All optimal circles C intersect only one fixed point

However, we have the following weaker incidence property.

Lemma 3.22. *Let $\| \cdot \|$ be a norm in $\mathbb{R}^n$ and assume that $S(X,r) \in \mathcal{G}$ is a minisum hypersphere. Then $S(X,r)$ intersects the convex hull of the fixed points in at least two points, i.e.,*

$$|S(X,r) \cap \mathrm{conv}(\mathcal{D})| \geq 2. \tag{3.1}$$

Furthermore, if this set of intersection points, denoted by I, is finite, it has to be a subset of the fixed points $\mathcal{D}$.

Proof. The set of intersection points I cannot be empty, see Corollary 3.20. If I is finite, then all points in I must be extreme points of the convex hull of $\mathcal{D}$. Hence, in this case we have $I \subseteq \mathcal{D}$. It remains to show that $|I| = 1$ is not possible. To this end let $S(X,r)$ be a minisum hypersphere and suppose that $I = \{A_m\}$ for some fixed point $A_m \in \mathcal{D}$. Then two cases are possible. Either $\mathcal{D}$ is inside or outside of $S(X,r)$, i.e.,

$I_- = \mathscr{D} \setminus I$ or $I_+ = \mathscr{D} \setminus I$. In both cases we can decrease and increase, respectively, the radius r and adjust the center X such that the objective value $f(S(X,r))$ decreases.   □

Note that for the Euclidean norm Lemma 3.22 follows from the incidence property stated in Theorem 2.18.

The following results shows that the dominance criterion for the Euclidean norm stated in Lemma 2.17 extends to arbitrary norms.

Lemma 3.23. *Let $\| \cdot \|$ be a norm in $\mathbb{R}^n$ and suppose a minisum hypersphere $S(X,r) \in \mathscr{G}$ exists. If*

$$\text{Test}_m := 2\omega_m - \sum_{m=1}^{M} \omega_m > 0$$

for some $1 \leq m \leq M$, then $S(X,r)$ has to intersect the fixed point A_m, i.e., $A_m \in S(X,r)$. If $\text{Test}_m = 0$, then there exists at least one optimal solution that intersects A_m.

Proof. Replace the Euclidean norm in the proof of Lemma 2.17 by an arbitrary norm $\| \cdot \|$.   □

3.6 Polyhedral Norms in the Plane

The goal of this section is to derive a geometric description for the set of all points $X \in \mathbb{R}^2$ that correspond to the center of a minisum circle $C(X,r)$. Throughout this section we denote the unit ball of the polyhedral norm $\| \cdot \|$ with B and refer to its fundamental directions by

$$\text{Ext}(B) = \{E_g : 1 \leq g \leq s\}.$$

In order to exclude degenerated bisectors, we assume that no straight line $\langle A_i, A_j \rangle$, $1 \leq i < j \leq M$, is parallel to a flat spot of the unit ball B. Then Theorem 1.48 ensures that all bisectors B_{ij}, $1 \leq i < j \leq M$, are uniquely defined piecewise linear curves. Since any problem may be slightly perturbed to guarantee this condition, the assumption is not limiting in a practical setting.

Let $\mathscr{L}_{\mathscr{D}}$ be the set of all *direction lines* (see Notation 1.32) through each of the fixed points $A_m \in \mathscr{D}$. The set $\mathscr{L}_{\mathscr{D}}$ induces a subdivision of the plane into vertices, edges, and faces. Some of the edges and faces are unbounded. In Computational Geometry this subdivision is called the *arrangement* induced by $\mathscr{L}_{\mathscr{D}}$, see, e.g., [BKOS00]. We refer to this arrangement by $\mathscr{A}(\mathscr{L}_{\mathscr{D}})$. For our purposes it is convenient to assume that each face of the arrangement $\mathscr{A}(\mathscr{L}_{\mathscr{D}})$ is closed. Figure 3.6 shows an example of an arrangement. As it can be seen in Fig. 3.6, each face of $\mathscr{A}(\mathscr{L}_{\mathscr{D}})$ is given as the intersection

$$\bigcap_{m=1}^{M} \{A_m\} + \Gamma_{g(m)}$$

where $g(m) \in \{1,\ldots,s\}$ for $1 \leq m \leq M$, and $\Gamma_{g(m)}$ is a cone generated by adjacent fundamental directions of norm $\|\cdot\|$, see Section 1.3.4. In particular, each face of $\mathscr{A}(\mathscr{L}_{\mathscr{D}})$ is convex. For polyhedral norms each face of $\mathscr{A}(\mathscr{L}_{\mathscr{D}})$ is an *elementary convex set* according to the concept introduced by Durier and Michelot (see [DM85] for further details). In the following it is important that the distance $\|X - A_m\|$ is affine linear in X on each face of the arrangement $\mathscr{A}(\mathscr{L}_{\mathscr{D}})$.

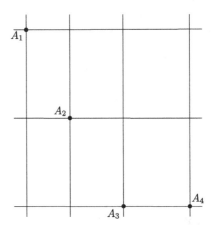

Fig. 3.6 Arrangement $\mathscr{A}(\mathscr{L}_{\mathscr{D}})$ induced by direction lines of the Manhattan norm through the fixed points A_1, A_2, A_3, and A_4. Each intersection point of two direction lines is a vertex of $\mathscr{A}(\mathscr{L}_{\mathscr{D}})$. Each line segment with end points at intersection points of direction lines and each ray with start point at an intersection point of two direction lines are edges of $\mathscr{A}(\mathscr{L}_{\mathscr{D}})$. Each two-dimensional cell is a face of $\mathscr{A}(\mathscr{L}_{\mathscr{D}})$

Lemma 3.24. *For any $A_{m'} \in \mathscr{D}$, the function $h_{m'}(X) = \|X - A_{m'}\|$ is affine linear in X on each face $\mathscr{F}$ of the arrangement $\mathscr{A}(\mathscr{L}_{\mathscr{D}})$.*

Proof. Let $\mathscr{F}$ be a face of the arrangement $\mathscr{A}(\mathscr{L}_{\mathscr{D}})$. Then, for each $A_m \in \mathscr{D}$ there exists $g(m) \in \{1,\ldots,s\}$ such that

$$\mathscr{F} = \bigcap_{m=1}^{M} \{A_m\} + \Gamma_{g(m)}.$$

Lemma 1.38 implies that $h_{m'}(X)$ is affine linear on the cone $\{A_{m'}\} + \Gamma_{g(m')}$. Since $\mathscr{F} \subseteq \{A_{m'}\} + \Gamma_{g(m')}$ we obtain the assertion. $\square$

From Lemma 3.24 we may conclude that the bisector $B_{ij} \cap \mathscr{F}$ has to be empty, a single point, a line segment, or a half-line, $1 \leq i < j \leq M$. (In general also $B_{ij} \cap \mathscr{F} = \mathscr{F}$ is possible, but our assumptions exclude this case.) Hence, we can refine the arrangement $\mathscr{A}(\mathscr{L}_{\mathscr{D}})$ by adding the set of bisectors $\{B_{ij} : 1 \leq i < j \leq M\}$ to $\mathscr{L}_{\mathscr{D}}$, i.e., we define

$$\mathscr{L} := \{B_{ij} : 1 \leq i < j \leq M\} \cup \mathscr{L}_{\mathscr{D}}$$

and conclude that $\mathscr{A}(\mathscr{L})$ is a refinement of $\mathscr{A}(\mathscr{L}_{\mathscr{D}})$. In particular, the faces of the arrangement $\mathscr{A}(\mathscr{L})$ remain convex. Each face of the arrangement $\mathscr{A}(\mathscr{L})$ is an *ordered elementary convex set* according to the concept introduced by Nickel and Puerto [NP05]. An example of an arrangement $\mathscr{A}(\mathscr{L})$ is depicted in Fig. 3.7.

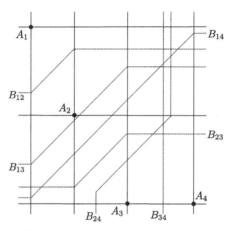

Fig. 3.7 Arrangement $\mathscr{A}(\mathscr{L})$ for the fixed points A_1, A_2, A_3, and A_4 where the Manhattan norm is used in order to define direction lines and bisectors $B_{ij}, 1 \leq i < j \leq 4$

Now, let $X \in \mathbb{R}^2$ be given and consider the one-dimensional problem

$$h_X(r) = f(C(X, r)) = \sum_{m=1}^{M} \omega_m d_m(C(X, r)) = \sum_{m=1}^{M} |\|X - A_m\| - r|.$$

Due to Corollary 3.2 we conclude that h_X is convex and piecewise linear in r. Hence, we may define

$$r_X^* = \min\{r \geq 0 : h_X(r) \leq h_X(r') \,\forall\, r' \geq 0\},$$

i.e., we take the smallest median if the median is not unique. We conclude that the circle $C(X, r_X^*)$ intersects at least one of the fixed points denoted as $A_{r_X^*}$. Given a face $\mathscr{F} \in \mathscr{A}(\mathscr{L})$ we now show that for all $X \in \mathscr{F}$ the circle $C(X, r_X^*)$ intersects the same unique fixed point $A_{r_X^*}$.

Lemma 3.25. *Let $\mathscr{F}$ be a face of the arrangement $\mathscr{A}(\mathscr{L})$. For all $X \in \mathscr{F}$, the circle $C(X, r_X^*)$ intersects the same unique fixed point $A_{r_X^*}$. Furthermore, the subsets J_-, J_0, and J_+ remain unchanged for all $C(X, r_X^*)$ with $X \in \mathscr{F}$.*

Proof. Consider any point $X \in \mathscr{F}$. Since X does not lie on a bisector, any circle $C(X, r)$ can intersect at most one fixed point. Thus, $A_{r_X^*}$ is unique for any given

$X \in \mathscr{F}$. That J_-, J_0, J_+ remain unchanged for all points in the interior of face $\mathscr{F}$ relies on the fact that for any change to occur, X must cross a bisector, which is impossible within $\mathscr{F}$. This means that the order of the distances $\|X - A_m\|$, $1 \leq m \leq M$, is the same for every point $X \in \mathscr{F}$. Consequently, for all $X \in \mathscr{F}$ a (smallest) median is attained at the same fixed point.   □

Given a face $\mathscr{F} \in \mathscr{A}(\mathscr{L})$ we denote as A_{r^*} the unique fixed point $A \in \mathscr{D}$ such that $A_{r^*} = A_{r_X^*}$ for all $X \in \mathscr{F}$.

Lemma 3.26. *Let $\mathscr{F}$ be a face of the arrangement $\mathscr{A}(\mathscr{L})$. For all $X \in \mathscr{F}$, the function $f(X, r^*(X))$ is affine linear in X where $r^*(X) = \|X - A_r^*\|$.*

Proof. For all $X \in \mathscr{F}$, using Lemma 3.25 the objective function may be rewritten as

$$f(X, r^*(X))$$
$$= \sum_{m:A_m \in J_-} \omega_m(\|X - A_{r^*}\| - \|X - A_m\|)$$
$$+ \sum_{m:A_m \in J_+} \omega_m(\|X - A_m\| - \|X - A_{r^*}\|).$$

where A_{r^*} and the subsets J_-, J_+ are fixed. Since $\mathscr{A}(\mathscr{L})$ is a refinement of $\mathscr{A}(\mathscr{L}_{\mathscr{D}})$ we obtain from Lemma 3.24 that within every face $\mathscr{F}$ $\|X - A_m\|$ is an affine linear function in X, $1 \leq m \leq M$. Thus, $f(X, r^*(X))$ is a sum of linear terms, and hence $f(X, r^*(X))$ is affine linear for all $X \in \mathscr{F}$.   □

The linearity of the function $f(X, r^*(X))$ within a face of $\mathscr{A}(\mathscr{L})$ leads to a geometric description of the set of minisum circles. In order to state this description we define

$$Q = \{C \in \mathscr{G} : f(C) \leq f(C') \ \forall \, C' \in \mathscr{G}\},$$
$$Q_X = \{X \in \mathbb{R}^2 : \exists \, r > 0 \text{ s.t. } C(X, r) \in Q\}.$$

The set Q_X together with Lemma 3.25 may be used in order to obtain an insight into the structure of the set Q. We show that a point $X \in Q_X$ cannot be contained in certain elements of the arrangement $\mathscr{A}(\mathscr{L})$. To this end, we introduce four categories for the vertices V of the arrangement $\mathscr{A}(\mathscr{L})$.

Type 1: V is the intersection of two direction lines.
Type 2: V is the intersection of a direction line and a bisector.
Type 3: V is the intersection of three bisectors B_{ij}, B_{ik}, and B_{jk} for pairwise distinct i, j, k.
Type 4: V is the intersection of two bisectors B_{ij} and B_{kl} for pairwise distinct i, j, k, l.

Note that a vertex of the arrangement $\mathscr{A}(\mathscr{L})$ may belong to different types. For instance, if a fixed point A_m is intersected by a bisector then A_m is a vertex of $\mathscr{A}(\mathscr{L})$

belonging to type 1 and type 2. To avoid such situations suppose that the fixed points in $\mathscr{D}$ fulfill the following conditions.

(i) For any triple L_i, L_j, L_k of distinct direction lines in $\mathscr{L}$, their intersection $L_i \cap L_j \cap L_k$ is either empty or is a fixed point $A_m \in \mathscr{D}$. Furthermore, no bisector intersects a fixed point $A_i \in \mathscr{D}$ and no direction line intersects two distinct fixed points $A_i, A_j \in \mathscr{D}$.

(ii) For two distinct direction lines $L_i, L_j \in \mathscr{L}$ and a bisector $B_{kl} \in \mathscr{L}$ their intersection $L_i \cap L_j \cap B_{kl}$ is empty. Analogously, for two distinct bisectors $B_{ij}, B_{kl} \in \mathscr{L}$ and a direction line $L_m \in \mathscr{L}$ their intersection $B_{ij} \cap B_{kl} \cap L_m$ is empty.

(iii) Three distinct bisectors $B_{i_1 i_2}, B_{i_3 i_4}, B_{i_5 i_6} \in \mathscr{L}$ have a common intersection point if and only if $\{i_k : 1 \leq k \leq 6\}$ contains exactly three distinct indices.

Conditions (i)–(iii) ensure that any vertex of the arrangement $\mathscr{A}(\mathscr{L})$ belongs exactly to one of the four types defined above. This allows to discard vertices of $\mathscr{A}(\mathscr{L})$ as centers of minisum circles. In this way we obtain a description of the set Q_X and therefore also for Q.

First, we show that a fixed point cannot be the center of a minisum circle.

Lemma 3.27. *Suppose that the fixed points in $\mathscr{D}$ fulfill conditions (i)–(iii) and let $C(X,r) \in \mathscr{G}$ be a minisum circle. Then we have $X \neq A$ for all $A \in \mathscr{D}$, i.e., the center of a minisum circle cannot be a fixed point.*

Proof. Assume that $A_s \in \mathscr{D}$ is the center of a minisum circle and delete the direction lines through A_s from $\mathscr{L}$. Then A_s becomes an internal point of some face $\mathscr{F}$ of the resulting arrangement $\mathscr{A}(\mathscr{L}')$. We conclude from Lemma 3.26 that the objective function less the contribution from A_s,

$$f_s(C(X,r^*(X))) = f(C(X,r^*(X))) - \omega_m d_m(C(X,r^*(X)))$$
$$= \sum_{m \neq s} \omega_m d_m(C(X,r^*(X))),$$

is affine linear within a δ-neighborhood of A_s where the radius δ is sufficiently small to ensure that the given neighborhood is contained entirely within face $\mathscr{F}$. Furthermore, $d_s(C(X,r^*(X))) = r^*(X) - \|X - A_s\|$ is strictly concave at A_s since $r^*(X)$ is affine linear and $\|Y\|$ is strictly convex at the origin. It follows that $f(X,r^*)$ is strictly concave at A_s, and hence, (A_s, r^*) cannot be the center of a minisum circle. $\quad\square$

As a consequence, we may conclude:

- Q_X does not contain a vertex of $\mathscr{A}(\mathscr{L})$ coinciding with a fixed point.
- Q_X does not contain points belonging to an edge of $\mathscr{A}(\mathscr{L})$ that contains fixed points.
- Q_X does not contain points belonging to the interior of a face of $\mathscr{A}(\mathscr{L})$ whose boundary contains fixed points.

Next, we show that Q_X has to contain more points than a vertex of $\mathscr{A}(\mathscr{L})$ corresponding to type 4.

Lemma 3.28. *Suppose that the fixed points in $\mathcal{D}$ fulfill conditions (i)–(iii) and let $C(X,r) \in \mathcal{G}$ be a minisum circle. If $X \in Q_X$ is the intersection of two bisectors B_{ij} and B_{kl} for pairwise distinct i,j,k,l, then Q_X has to contain at least all points belonging to two different edges of $\mathcal{A}(\mathcal{L})$ which have an end point in X.*

Proof. Assume that $C(X,r)$ is a minimum circle and $X = B_{ij} \cap B_{kl}$ where i,j,k,l are pairwise distinct. Conditions (i)–(iii) imply that no other lines in $\mathcal{L}$ than B_{ij} and B_{kl} intersect X. Therefore it follows that $C^* = C(X,r^*(X))$ may contain at most two fixed points, A_i, A_j or A_k, A_l. Without loss of generality assume that $A_i, A_j \in C^*$. Then A_k, A_l are either in the interior or exterior of C^*. In both cases the objective function $f(C(X,r^*(X)))$ has to be affine linear on the segment of B_{ij} joining the adjacent vertices of $\mathcal{A}(\mathcal{L})$ on either side of X. Therefore $X \in Q_X$ cannot be a unique minimizer of $f(C(X,r^*(X)))$. Hence, all points contained in the edges $\mathcal{E}_1, \mathcal{E}_2 \subseteq B_{ij}$ of $\mathcal{A}(\mathcal{L})$ having their end points in X have to belong to Q_X.   □

Remark 3.29. Besides the results presented in this section, [BJKS09] contains further results on the four different types of vertices of the arrangement $\mathcal{A}(\mathcal{L})$.

1. Let X be any vertex of type 1 other than a fixed point. X corresponds to two distinct fixed points, say A_1 and A_2. If one of these points lies within the circle $C^* = C(X,r^*(X))$, then C^* is not a minisum circle and $X \notin Q_X$.
2. Let X be any vertex of $\mathcal{A}(\mathcal{L})$ belonging to type 2 or type 3. If the circle $C(X,r^*(X))$ is a minisum circle that does not intersect the pair A_i, A_j associated with type 2 or the triplet A_i, A_j, A_k associated with type 3, then Q_X has to contain all points of an edge of $\mathcal{A}(\mathcal{L})$ that has an end point in X.
3. Let X be any vertex belonging to type 2 and assume that the circle $C^* = C(X,r^*(X))$ is a minisum circle. If the fixed point associated with the single direction line through X lies inside C^*, then Q_X has to contain all points of an edge of $\mathcal{A}(\mathcal{L})$ that has an end point in X.

Using the results of this section it is possible to construct a solution method based on an enumeration of the vertices of type 1, 2, and 3 of the arrangement $\mathcal{A}(\mathcal{L})$. In order to state an algorithm which is based on this idea we denote the direction line through A_m and parallel to the fundamental direction E_g as L_{mg}, $1 \leq m \leq M$, $1 \leq g \leq s$. Furthermore, we assume that the fundamental directions of the norm $\| \cdot \|$ are numbered from 1 to s in clockwise direction. Recall that this implies that direction line L_{mg} and direction line $L_{m(g+s/2)}$ coincide for all $g = 1,\ldots,s/2$ (see Section 1.3.4). With these conventions we can formulate Algorithm 3.1 which computes a minisum circle under the assumption that the set of fixed points $\mathcal{D}$ fulfills conditions (i)–(iii).

Theorem 3.30. *Let $\| \cdot \|$ be a polyhedral norm in $\mathbb{R}^2$ with s fundamental directions and suppose that the set of fixed points $\mathcal{D}$ fulfills conditions (i)–(iii). Then Algorithm 3.1 computes a minisum circle in $\mathcal{O}(s^2 M^4)$ operations.*

Proof. Note that for a vertex X of type 2 or 3 an optimal radius r^* need not be computed. Indeed, using the second result stated in Remark 3.29 it is sufficient

---

Algorithm 3.1: Computing a minisum circle under a polyhedral norm

---

Input: Polyhedral norm $\|\cdot\|$ with unit ball B, set of fixed points $\mathscr{D}$ satisfying conditions
  (i)–(iii), associated weights ω_m
Output: Minisum circle $C(X_{opt}, r_{opt})$
// Initialization
1 $X := X_{opt} := (0,0)$, $r_{opt} := 1$, $\mathscr{I} = \emptyset$
// Computation of bisectors
2 **for** $1 \leq i < j \leq M$ **do**
3 $\quad \lfloor$ Compute B_{ij}

// Type 1 vertices
4 **for** $1 \leq i < j \leq M$, $1 \leq g < h \leq s/2$ **do**
5 $\quad \mid X := (L_{ig} \cap L_{jh})$
6 $\quad \mid$ Compute $r \in \text{median}\{\omega_m \|X - A_m\| : 1 \leq m \leq M\}$
7 $\quad \mid$ **if** $f(C(X,r)) < f(C(X_{opt}, r_{opt}))$ **then**
8 $\quad \lfloor \quad \lfloor X_{opt} := X$, $r_{opt} := r$

// Type 2 vertices
9 **for** $1 \leq m \leq M$, $1 \leq g \leq s/2$, $1 \leq i < j \leq M$ **do**
10 $\quad \mid \mathscr{I} := L_{mg} \cap B_{ij}$
11 $\quad \mid$ **for** $X \in \mathscr{I}$ **do**
12 $\quad \mid \quad \mid r := \|X - A_m\|$
13 $\quad \mid \quad \mid$ **if** $f(C(X,r)) < f(C(X_{opt}, r_{opt}))$ **then**
14 $\quad \lfloor \quad \lfloor \quad \lfloor X_{opt} := X$, $r_{opt} := r$

// Type 3 vertices
15 **for** $1 \leq i < j < k \leq M$ **do**
16 $\quad \mid X := \text{Ext}(B_{ij} \cap B_{ik})$ $r := \|X - A_m\|$
17 $\quad \mid$ **if** $f(C(X,r)) < f(C(X_{opt}, r_{opt}))$ **then**
18 $\quad \lfloor \quad \lfloor X_{opt} := X$, $r_{opt} := r$

---

to set $r^* = \|X - A_m\|$ where A_m is a fixed point associated with X. Therefore it follows that at least one minisum circle is contained in the set of circles evaluated by Algorithm 3.1.

The algorithm has to compute $\mathscr{O}(M^2)$ bisectors. For this purpose an optimal algorithm proposed by Icking et al. [IKM$^+$99] may be used which computes a bisector in $\mathscr{O}(s)$ operations. Thus the computation of all bisectors need $\mathscr{O}(sM^2)$ operations.

Since there are $\mathscr{O}(sM)$ direction lines, we have $\mathscr{O}(s^2M^2)$ type 1 vertices. Given a vertex of type 1, $\mathscr{O}(M \log M)$ operations are required to determine an optimal radius r^* and again $\mathscr{O}(M)$ operations are needed to determine the objective value of the circle $C(X, r^*)$. Due to conditions (i)–(iii) each vertex of type 1 is the unique intersection point of two direction lines which can be computed in constant time. Thus, the vertices of type 1 may be evaluated in $\mathscr{O}(s^2M^3 \log M)$ operations. As mentioned above, we have $\mathscr{O}(sM)$ direction lines and $\mathscr{O}(M^2)$ bisectors. Since each bisector contains at most $4s$ linear pieces, see Theorem 1.54, we have $\mathscr{O}(s^2M^3)$ vertices of type 2. Again, each vertex is the unique intersection between a direction line and a linear piece of a bisector and may be computed in constant time. Thus, the vertices of type 2 may be evaluated in $\mathscr{O}(s^2M^4)$ operations. Finally, we have $\mathscr{O}(M^3)$

type 3 vertices. Analog to type 2 vertices, we need $\mathscr{O}(M^4)$ operations to evaluate all circles corresponding to this type of vertices. Note that conditions (i)–(iii) do not ensure that the intersection of two distinct bisectors is a point. It is possible that such an intersection is a half-line and therefore we have to choose the unique extreme point of the intersection between two bisectors B_{ij}, B_{ik} in order to obtain vertices of type 3. Nevertheless, also this can be done in constant time. Summing up, $\mathscr{O}(s^2 M^4)$ operations are required by Algorithm 3.1.   □

Remark 3.31. Note that Algorithm 3.1 is also valid if the fixed points in $\mathscr{D}$ do not fulfill conditions (i)–(iii). But in that case some of the intersections computed in the algorithm may become line segments, half-lines, or straight line. Thus, the algorithm has to be adapted slightly to be able to handle these cases.

3.7 Concluding Remarks

Several results for the minisum hypersphere problem with Euclidean norm also apply for the problem with an arbitrary norm. For instance, the radius of a minisum hypersphere is always greater than zero. Furthermore, if a minisum hypersphere $S \in \mathscr{G}$ exists, then also a minisum hypersphere exists that intersects at least one fixed point. However, an optimal solution to the minisum hypersphere problem with an arbitrary norm need not include two fixed points; this makes the problem rather more difficult to solve than the standard Euclidean case and the solution approaches described in Section 2.6 cannot be applied. Even for the planar case $n = 2$ a general solution approach for arbitrary norms is not available. A handicap for the development of a solution approach is the absence of minisum hyperspheres for any problem instance. Thus, an algorithm has to test whether or not the problem is unbounded in the sense that increasing the radius of a hypersphere may result in a superior hypersphere. For the polyhedral case this behavior of the objective function is also possible, however here it can be shown that a minisum circle having finite radius always exists. For arbitrary norms the problem of nonexistence of minisum hyperspheres may be avoided by defining an upper bound for the radius of a minisum hypersphere. Besides this rather small problem, the main obstacle for a feasible algorithm for all norms is the lack of a simple description of the unit ball of a norm in terms of its coordinates in $\mathbb{R}^n$. For instance, for polyhedral norms, L_p norms, or elliptic norms such a description is known. Therefore for theses classes of norms general approaches of global optimization may be applied. The big cube small cube method was already applied in [Sch07] in order to solve the minisum hypersphere problem with Euclidean norm L_2 and Manhattan norm L_1. The transfer to other L_p norms seems to be readily possible. The main idea of the big cube small cube method also leads to algorithms for other norms, provided that the norm can be evaluated in any point and lower and upper bounds are available. But from a geometric point of view the big cube small cube approach is not of interest since it only approximates a minisum hypersphere and gives no insight into the structure of the problem.

   Besides the lack of a solution method, also other interesting points are still open. For instance, is there a sensitivity of a minisum hypersphere to perturbations of the fixed points? For the case of squared Euclidean distance this question was studied in [Nie04]. It could also be of interest to study heuristics for the minisum hypersphere problem. For smooth and strictly convex norms a hypersphere in $\mathbb{R}^n$ is uniquely described by $n+1$ points, provided that these points are not collinear. Hence, an interesting heuristic which should be analyzed more deeply consists of choosing the best hypersphere among all hyperspheres defined by $n+1$ distinct fixed points.

Chapter 4
Minisum Circle Problem with Unequal Norms

4.1 Basic Assumptions

This chapter considers an extension of the minisum circle problem which may be described in the following way:

Dimensions: $n = 2, M \geq 2$
Norm: Arbitrary norm $\|\cdot\|$ in $\mathbb{R}^2$
Objects: Circles defined with respect to $\|\cdot\|$
Metric: Induced by a norm k in $\mathbb{R}^2$; $d(X,Y) := k(Y - X)$
Distance: $d(S,A_m) = \min\{d(Y,A_m) : Y \in S\}, 1 \leq m \leq M$
Objective: $f(S) = \sum_{m=1}^{M} \omega_m d(S,A_m) \to \min$

The problem considered in this chapter is called *minisum circle problem with unequal norms*. Note that the distance between a circle and the fixed points is measured by a metric induced by a second norm k, i.e., two norms are used in this chapter: norm $\|\cdot\|$ is used in order to define the set of circles $\mathscr{G} \subseteq \mathbb{R}^2$ and norm k is used in order to define the point–circle distance. The class of norms considered in this chapter is mostly the class of polyhedral norms; that is, $\|\cdot\|$ and k are both polyhedral norms.

Besides the applications of circle location models mentioned in Section 1.2.2, the minisum circle problem with unequal norms includes the design of circular public transportation networks. Circular public transportation networks are common in practice. For instance, in London, Moscow, Berlin, Hamburg, and Tokyo, circular underground or suburban railways can be found. An optimal solution to the minisum circle problem with unequal norms is suited to determine a rough route of a new circular public transportation network that minimizes the distances from the customers to the public transportation network. In a subsequent detailed planning, this tentative route can be adapted to local realities (e.g., buildings, watercourses, parks, and

M.-C. Körner, *Minisum Hyperspheres*, Springer Optimization and Its Applications 51,
DOI 10.1007/978-1-4419-9807-1_4, © Springer Science+Business Media, LLC 2011

nature protection areas). Similarly, the minisum circle problem with unequal norms may be used to determine ring roads which may also be of practical interest, see Fujita and Suzuki [FS04] and Pearce [Pea74].

It should be noted that this chapter is based on [KBJS10] which is a joint work with Brimberg, Juel, Schöbel, and the author.

4.2 Distance

In this section we study the point–circle distance $d(C,A)$. As mentioned above, the distance between a circle $C(X,r)$ and a fixed point $A_m \in \mathscr{D} \subseteq \mathbb{R}^2$ is measured in the metric induced by the norm k; thus we have

$$d(C(X,r),A_m) = \min_{Y \in C(X,r)} d(A_m,Y)$$

$$= \min\{k(A_m - Y) : \|Y - X\| = r, Y \in \mathbb{R}^2\}. \tag{4.1}$$

As before, the shortcut $d_m(C(X,r))$ is used in order to refer to distance between the circle $C(X,r)$ and the fixed point A_m. Note that the point–circle distance defined in (4.1) may differ strongly from the point–circle distance used in previous chapters whenever $k \neq \|\cdot\|$. However, in the following it is shown that some properties of the point–circle distance are not affected by allowing a second norm k. We start with the extension of the symmetry property which is obvious in case $k = \|\cdot\|$, see Corollary 3.4.

Lemma 4.1. *Let k and $\|\cdot\|$ be arbitrary norms in $\mathbb{R}^2$. Given two points $A,X \in \mathbb{R}^2$ and $r > 0$ we have $d(C(X,r),A) = d(C(A,r),X)$.*

Proof. Let $C_1 = C(X,r)$ and $C_2 = C(A,r)$. Let $Y_1 \in C_1$ such that $d(C_1,A) = k(Y_1 - A)$. Let $Y_2 := A + X - Y_1$. Note that $Y_2 \in C_2$, since $\|Y_2 - A\| = \|X - Y_1\| = r$. We have

$$k(Y_2 - X) = k(A + X - Y_1 - X) = k(A - Y_1) = d(C_1,A);$$

hence, $d(C_2,X) \leq d(C_1,A)$. Now, we choose $Z_1 \in C_2$ such that $d(C_2,X) = k(Z_1 - X)$. Analogous to the previous case we obtain $d(C_2,X) \geq d(C_1,A)$; that is, $d(C_1,A) = d(C_2,X)$. □

As in Chapter 3 it can be shown that the point–circle distance $d(C(X,r),A)$ behaves on certain regions like a concave and a convex function, respectively; that is, the convexity and concavity properties are not affected by the norm k. In order to prove this property and also other results for the minisum circle problem with different norms it is helpful to use points $(X,r) \in \mathbb{R}^2 \times]0,\infty[$ to identify circles $C(X,r) \in \mathscr{G}$. Further, we need the following auxiliary lemma.

Lemma 4.2. *Let A be a fixed point and $H \subseteq \mathbb{R}^3$ be any supporting plane of the set*

$$V := \{(X,r) \in \mathbb{R}^2 \times [0,\infty[: \|X - A\| \leq r\}.$$

Then the distance

$$d(X, r, H) = \min\{k(X - Y) : (Y, r) \in H\}$$

between the supporting plane H and the point $(X, r) \in V$ is a linear function of (X, r) in V.

Proof. Let us assume that $\alpha \in \mathbb{R}^3$, $\alpha \neq 0$, and $\beta \in \mathbb{R}$ such that

$$H = \left\{ \begin{pmatrix} y_1 \\ y_2 \\ y_3 \end{pmatrix} \in \mathbb{R}^3 : \alpha_1 y_1 + \alpha_2 y_2 + \alpha_3 y_3 = \beta \right\}.$$

The set

$$H(r) := \{Y \in \mathbb{R}^2 : (Y, r) \in H\}$$

$$= \left\{ \begin{pmatrix} y_1 \\ y_2 \end{pmatrix} \in \mathbb{R}^2 : \alpha_1 y_1 + \alpha_2 y_2 = \beta - \alpha_3 r \right\}$$

is a straight line in $\mathbb{R}^2$. Hence,

$$d(X, r, H) = \min\{k(X - Y) : Y \in H(r)\}$$

is the k-distance between the point $X \in \mathbb{R}^2$ and the straight line $H(r) \subseteq \mathbb{R}^2$. For the distance between a point $X = (x_1, x_2)$ and a straight line $H(r)$ we can apply the formula stated in Corollary 1.1 in [PC01] and obtain

$$d(X, r, H) = \min\{k(X - Y) : Y \in H(r)\}$$
$$= \frac{|\beta - \alpha_3 r - \alpha_1 x_1 - \alpha_2 x_2|}{k^\circ(\overline{\alpha})}$$

where $\overline{\alpha} := (\alpha_1, \alpha_2)$ and k° is the dual norm of k. Since the denominator $k^\circ(\overline{\alpha})$ does not depend on (X, r) it follows that $d(X, r, H)$ is linear in (X, r) on both half-spaces defined by H. With the fact that V is included in a half-space defined by the supporting plane H we obtain the assertion.   □

Lemma 4.3. *Let k and $\|\cdot\|$ be arbitrary norms in $\mathbb{R}^2$. Then the point–circle distance $d(C(X, r), A)$ between a circle $C(X, r)$ and a fixed point A is*

(i) concave in (X, r) on the set

$$V := \{(X, r) \in \mathbb{R}^2 \times [0, \infty[: \|X - A\| \leq r\}$$

   and
(ii) convex in (X, r) on any convex set $U \subseteq (\mathbb{R}^2 \times]0, \infty[) \setminus V$.

Proof. We start with the proof of convexity of the point–circle distance; afterward, we prove concavity.

Convexity. Let $U \subseteq (\mathbb{R}^2 \times]0, \infty[) \setminus V$ be a convex set. Consider any two points $(X_1, r_1), (X_2, r_2) \in U$. Let $X_3 := \lambda X_1 + (1 - \lambda)X_2$ and $r_3 := \lambda r_1 + (1 - \lambda)r_2$ for some $\lambda \in [0, 1]$. Using Lemma 4.1, we get

$$d(C(X_i, r_i), A) = d(C(A, r_i), X_i) = k(X_i - Z_i)$$

where $Z_i \in C(A, r_i)$ minimizes the k-distance from X_i to $C(A, r_i)$, $i = 1, 2, 3$. We have

$$\|\lambda Z_1 + (1 - \lambda)Z_2 - A\| \leq \lambda \|Z_1 - A\| + (1 - \lambda)\|Z_2 - A\| = r_3;$$

that is, we may conclude that $\lambda Z_1 + (1 - \lambda)Z_2$ is either on $C(A, r_3)$ or in its interior. Since X_3 does not belong to $C(A, r_3)$ or its interior we obtain

$$
\begin{aligned}
k(X_3 - Z_3) &= \min\{k(X_3 - Z) : Z \in C(A, r_3)\} \\
&= \min\{k(X_3 - Z) : \|Z - A\| \leq r_3\} \\
&\leq k(X_3 - (\lambda Z_1 + (1 - \lambda)Z_2)) \\
&= k(\lambda X_1 + (1 - \lambda)X_2 - (\lambda Z_1 + (1 - \lambda)Z_2)).
\end{aligned}
$$

Thus

$$
\begin{aligned}
d(C(X_3, r_3), A) &= d(C(A, r_3), X_3) \\
&= k(X_3 - Z_3) \\
&\leq k(\lambda(X_1 - Z_1) + (1 - \lambda)(X_2 - Z_2)) \\
&\leq \lambda k(X_1 - Z_1) + (1 - \lambda)k(X_2 - Z_2) \\
&= \lambda d(C(A, r_1), X_1) + (1 - \lambda)d(C(A, r_2), X_2) \\
&= \lambda d(C(X_1, r_1), A) + (1 - \lambda)d(C(X_2, r_2), A).
\end{aligned}
$$

Concavity. Let $\partial V := V \setminus \text{int}(V)$ denote the boundary of V. For any point $(X, r) \in \partial V, r > 0$, let $H(X, r)$ denote a supporting plane of V at (X, r). $H(X, r)$ exists due to the convexity of the cone V and contains the ray from apex $(A, 0)$ passing through (X, r). (Note: if $\|\cdot\|$ is a smooth norm then $H(X, r)$ is uniquely defined. If $\|\cdot\|$ has corners then infinitely many supporting planes will occur at each corner, in which case anyone may be selected arbitrarily.) Furthermore, we define

$$\mathscr{H} := \{H(X, r) : (X, r) \in \partial V \setminus (A, 0)\}.$$

For any plane $H \in \mathscr{H}$ and any point $(X, r) \in V$ we define the distance

$$d(X, r, H) := \min\{k(X - Y) : (Y, r) \in H\}.$$

. Now, we consider an arbitrary point $(X, r) \in V$. Let $Y \in C(A, r)$ such that

$$k(X - Y) = d(C(A, r), X) = d(C(X, r), A).$$

Due to the definition of $\mathscr{H}$ we have

$$d(C(X,r),A) = k(X-Y) \leq d(X,r,H) \; \forall \, H \in \mathscr{H}.$$

Furthermore, specifically for $H(Y,r)$ we have $(Y,r) \in H$ and therefore

$$d(C(X,r),A) = k(X-Y) = d(X,r,H).$$

It follows that

$$d(C(X,r),A) = \min\{d(X,r,H) : H \in \mathscr{H}\} \; \forall \, (X,r) \in V.$$

But the distance $d(X,r,H)$ between (X,r) and a supporting plane H is a linear function of (X,r) in V, as it is shown in Lemma 4.2. Thus, $d(C(X,r),A)$ is the pointwise minimum of a (infinite) set of linear functions and is itself concave for all $(X,r) \in V$. $\qquad\square$

Now, we show that the point A is a local maximum of $h_r(X) := d(C(X,r),A)$. Note that this result proves that the center X of a circle $C(X,r)$ (defined w.r.t. norm $\|\cdot\|$) is a point within $C(X,r)$ which has the greatest k-distance to $C(X,r)$. This property may also apply to other points within the circle; that is, the stronger formulation $h_r(A) > h_r(X)$ is not generally true. Again, this result is obviously in case $k = \|\cdot\|$.

Lemma 4.4. *Let k and $\|\cdot\|$ be arbitrary norms in $\mathbb{R}^2$. Given a point $A \in \mathbb{R}^2$ and $r > 0$ define $h_r(X) = d(C(X,r),A)$. We have $h_r(A) \geq h_r(X)$ for all $X \in B(A,r) = \{Y : \|Y-A\| \leq r\}$.*

Proof. First note that

$$h_r(X) = d(C(A,r),X) = \min_{Y \in C(A,r)} k(X-Y)$$

is concave in X on the ball $B(A,r)$, see Lemma 4.1 and Lemma 4.3. Let us assume that there exists $Z \in B$ such that $h_r(Z) > h_r(A)$. Let $P \in C(A,r)$ such that $h_r(Z) = k(Z-P)$. We define

$$Z' := A - (Z-A),$$
$$P' := A - (P-A).$$

Then we have $\|A-P'\| = \|P-A\| = r$; that is, $P' \in C(A,r)$. Furthermore we have $k(Z'-P') = k(Z-P)$. We obtain

$$h_r(Z') \leq k(Z'-P') = k(Z-P) = h_r(Z).$$

Now, let $Q' \in C(A, r)$ such that $h_r(Z') = k(Z' - Q')$ and define $Q := A - (Q' - A)$. Analogously to the previous case we obtain $h_r(Z) \leq h_r(Z')$; hence, $h_r(Z) = h_r(Z')$. Since $A = \frac{1}{2}(Z + Z')$ we end up in a contradiction to the concavity of h_r:

$$h_r\left(\frac{1}{2}(Z + Z')\right) = h_r(A) < h_r(Z) = \frac{1}{2}h_r(Z) + \frac{1}{2}h_r(Z').$$

$\square$

In Chapter 3 the following simple formula for the point–circle distance between a point A and a circle $C = C(X, r)$ is stated:

$$d(C, A) = |k(X - A) - r|.$$

Obviously, this formula is not true if $\|\cdot\| \neq k$ and the point–circle distance has to be calculated using a numerical method. However, we are able to state simple lower and upper bounds for the point–circle distance using the norm $\|\cdot\|$. To this end we define

$$K := \max\{k(X) : \|X\| = 1\}, \qquad G := \min\{k(X) : \|X\| = 1\},$$

i.e., for all $A, X \in \mathbb{R}^2$, $A \neq X$, we have

$$G \leq \frac{k(A - X)}{\|A - X\|} \leq K.$$

For instance, with $\|\cdot\| = \|\cdot\|_1$, $k = \|\cdot\|_2$ we obtain that $K = 1$ and $G = \frac{1}{2}\sqrt{2}$. With these notations we obtain the following upper bound.

Lemma 4.5. *Let k and $\|\cdot\|$ be arbitrary norms in $\mathbb{R}^2$ and let $A \neq X$. Then*

$$d(C(X, r), A) \leq K |\|A - X\| - r|.$$

Proof. Let $\lambda \geq 0$ such that $A' := X + \lambda(A - X) \in C(X, r)$. We obtain

$$\|X - A'\| = r \Leftrightarrow \lambda \|A - X\| = r \Leftrightarrow \lambda = \frac{r}{\|A - X\|}.$$

With this fact we get

$$d(C(X, r), A) \leq k(A - A') = k(A - X - \lambda(A - X))$$

$$= |1 - \lambda| k(A - X) = \left|1 - \frac{r}{\|A - X\|}\right| k(A - X)$$

$$= \frac{k(A - X)}{\|A - X\|} |\|A - X\| - r| = K |\|A - X\| - r|.$$

$\square$

The following lemma states a lower bound for the point–circle distance.

Lemma 4.6. *Let k and $\|\cdot\|$ be arbitrary norms in $\mathbb{R}^2$ and let $A \neq X$. Then we have*

$$d(C(X,r),A) \geq \begin{cases} k(A-X) - Kr & \text{if} \quad \|A-X\| \geq r \\ Gr - k(A-X) & \text{if} \quad \|A-X\| \leq r \end{cases}.$$

Proof. Let $A' \in C(X,r)$ such that $d(C(X,r),A) = k(A-A')$.
If $\|A-X\| \geq r$ then we have

$$
\begin{aligned}
k(A-X) &\leq k(A-A') + k(A'-X) \\
&= d(C(X,r),A) + k(A'-X) \\
&\leq d(C(X,r),A) + K\|A'-X\| \\
&= d(C(X,r),A) + Kr.
\end{aligned}
$$

If $\|A-X\| \leq r$ then we have

$$k(A'-X) \leq k(A-A') + k(A-X) = d(C(X,r),A) + k(A-X);$$

hence, we obtain

$$
\begin{aligned}
d(C(X,r),A) &\geq k(A'-X) - k(A-X) \\
&\geq G\|A'-X\| - k(A-X) \\
&= Gr - k(A-X).
\end{aligned}
$$

$\square$

4.3 Properties of Minisum Circles

In this section we study properties of optimal solutions to the minisum circle problem with unequal norms $\|\cdot\|$ and k. In particular, we discuss properties that are known for the Euclidean case $\|\cdot\| = k = \|\cdot\|_2$ and the equal norms case $\|\cdot\| = k$ but do not apply to the case $\|\cdot\| \neq k$. We start our analysis with a property that extends the equal norm case $\|\cdot\| = k$ to the current case.

Lemma 4.7. *Let k and $\|\cdot\|$ be arbitrary norms in $\mathbb{R}^2$ and assume that $\mathscr{G}$ contains a minisum circle $C(X,r)$. Then the radius of r has to be positive.*

Proof. Assume that a circle $C(X,r)$ with radius $r = 0$ is an (degenerated) optimal solution and construct a circle C' that intersects X and a fixed point $A \neq X$. Analogously to Lemma 2.5 we obtain a contradiction. $\square$

From Lemma 4.7 it follows that a point can never be a degenerated minisum circle, provided that $M > 1$. Recall that another extreme is possible. As we have seen in Chapter 3 the minisum circle problem with equal norms is unbounded in the sense that increasing the radius may continuously improve the objective value. Due to the fact that this problem is a special case of the minisum circle problem with unequal norms, it is possible that a minimizer for the present problem is not contained in $\mathscr{G}$. In this case for instance a straight line or a bent line may solve the problem.

Remark 4.8. For the case of two polyhedral norms $\| \cdot \|$ and k in $\mathbb{R}^2$ it will become clear in Section 4.4 that $\mathscr{G}$ always contains a minisum circle. Recall that for other norms in Chapter 3 a restricted set $\mathscr{G}_{\bar{r}}$ of feasible hyperspheres is introduced in order to ensure existence of restricted minisum hyperspheres. An analog approach for the current problem is possible.

Recall that for the case of equal norms $\| \cdot \| = k$ and under the assumption of the existence of a minisum hypersphere in $\mathscr{G}$ a minisum hypersphere exists which intersects at least one fixed point, see Corollary 3.16. It is not possible to extend this incidence property to the case of unequal norms. This negative result is shown by the following example.

Example 4.9. Let $\| \cdot \| = \ell_\infty$ and $k = \| \cdot \|_1$, and consider

$$A_1 = (1,2), \qquad A_2 = (8,15), \qquad A_3 = (-2,16),$$
$$A_4 = (0,17), \qquad A_5 = (16,20)$$

with weights $\omega_1 = 1.1$, $\omega_2 = 2.2$, $\omega_3 = \omega_4 = 1$, and $\omega_5 = 1.05$. Using the results of Section 4.4 we obtain that any optimal circle minimizing the weighted sum of distances to the fixed points is contained in the set

$$\text{Opt} := \{C(X,r) : X = (8,8) + (r-8)(1,-1), \ r \geq 8\}.$$

Any optimal circle $C \in \text{Opt}$ has an objective function value of

$$1.1 \cdot 1 + 2.2 \cdot 1 + 2 + 1 + 1.05 \cdot 4 = 10.5.$$

In particular, as it may be read from Fig. 4.1, none of the optimal circles intersects any of the existing points.

Note that the minisum circle $C((8,8),8)$ in Example 4.9 does not satisfy the median property. Thus, a further consequence of Example 4.9 is that in contrast to the case of equal norms $\| \cdot \| = k$, see Corollary 3.18, a minisum circle for the minisum circle problem with unequal norms need not satisfy this property.

We close this section with a positive result. The dominance criterion for the minisum hypersphere problem with equal norms, see Lemma 3.23, also extends to the minisum circle problem with unequal norms.

Lemma 4.10. *Let $\| \cdot \|$ and k be two norms in $\mathbb{R}^2$ and suppose a minisum circle $C(X,r) \in \mathscr{G}$ exists. If*

$$\text{Test}_m := 2\omega_m - \sum_{m=1}^{M} \omega_m > 0$$

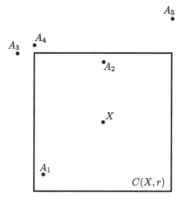

Fig. 4.1 Illustration of Example 4.9 with fixed points $A_1, \ldots, A_5$ and a minisum circle $C(X,r)$ where $X = (8,8)$ and $r = 8$

for some $1 \leq m \leq M$, *then* $C(X,r)$ *must intersect the fixed point* A_m. *If* $\text{Test}_m = 0$, *then there exists at least one optimal solution that intersects* A_m.

Proof. Analog to the proof of Lemma 3.23.   □

4.4 Polyhedral Norms

In this section we analyze the case in which both norms $\|\cdot\|$ and k are polyhedral norms; that is, for $X \in \mathbb{R}^2$

$$\|X\| = \min \left\{ \sum_{g=1}^{s} |\beta_g| : \sum_{g=1}^{s} \beta_g E_g = X \right\},$$

$$k(X) = \min \left\{ \sum_{g=1}^{t} |\beta_g| : \sum_{g=1}^{t} \beta_g F_g = X \right\},$$

where $E_g, g = 1, \ldots, s$, and $F_g, g = 1, \ldots, t$, are, respectively, the fundamental directions of the given polyhedral norms $\|\cdot\|$ and k. We refer to the set of fundamental directions of norm $\|\cdot\|$ by $\text{Ext}(B_1)$ and to the fundamental directions of k by $\text{Ext}(B_2)$; that is, the unit balls of the norms $\|\cdot\|$ and k are given by B_1 and B_2, respectively. Throughout this section we assume that the fundamental directions of $\|\cdot\|$ and k are numbered in clockwise order. For any fundamental direction E_i of norm $\|\cdot\|$ we denote the fundamental direction adjacent to E_i in clockwise direction as E_i^+; that is, $E_i^+ = E_{i+1}$ for $1 \leq i \leq s-1$ and $E_s^+ = E_1$. We define F_i^+ analogously; that is, for a fundamental direction F_i of norm k we have $F_i^+ = F_{i+1}$ for $1 \leq i \leq t-1$ and $F_t^+ = F_1$.

In contrast to the situation in Section 3.6 we consider *arrangements* of $\mathbb{R}^3$. In order to define an arrangement in $\mathbb{R}^3$, let $\mathcal{H}$ be a set of hyperplanes in $\mathbb{R}^3$. The hyperplanes in $\mathcal{H}$ define a partition of $\mathbb{R}^3$ into vertices, edges, two-dimensional cells, and three-dimensional cells. We refer to this subdivision as the *arrangement* induced by $\mathcal{H}$ and use the notation $\mathscr{A}(\mathcal{H})$. For sake of simplicity we refer to an element $\mathscr{F}$ of $\mathscr{A}(\mathcal{H})$ by *l-face*, provided that the smallest affine subspace of $\mathbb{R}^3$ containing $\mathscr{F}$ has dimension l, $0 \leq l \leq 3$. We also use the denotation *vertex* in order to refer to a 0-face of $\mathscr{A}(\mathcal{H})$. Note that l-faces are convex polytopes and may be unbounded. We are mainly interested in l-faces of an arrangement $\mathscr{A}(\mathcal{H})$ which are contained in $\mathbb{R}^2 \times [0, \infty[$. Therefore we refer to $\mathscr{A}^+(\mathcal{H})$ as the arrangement which we obtain from $\mathscr{A}(\mathcal{H})$ by adding the hyperplane $H = \{(x_1, x_2, x_3) \in \mathbb{R}^3 : x_3 = 0\}$ to $\mathcal{H}$ and then removing all l-faces contained in $\mathbb{R}^2 \times]-\infty, 0[$. For our purposes it is convenient to assume that each 3-face of the arrangement $\mathscr{A}^+(\mathcal{H})$ is closed.

In the following we identify a set of hyperplanes in $\mathbb{R}^3$ such that the objective function of the minisum circle problem with unequal polyhedral norms is concave on each 3-face of the arrangement $\mathscr{A}^+(\mathcal{H})$ induced by $\mathcal{H}$. In order to obtain the arrangement $\mathscr{A}^+(\mathcal{H})$, we first identify a set of hyperplanes $\mathcal{H}_m$ such that the point–circle distance $d(C(X, r), A_m)$ is concave in X and r on any 3-face of the arrangement $\mathscr{A}^+(\mathcal{H}_m)$. Subsequently, we will see that $A^+(\mathcal{H})$ can be obtained by combining the arrangements $\mathscr{A}^+(\mathcal{H}_m)$, $1 \leq m \leq M$. Furthermore, it will turn out that $A^+(\mathcal{H})$ gives rise to a finite dominating set according to Definition 1.56. It should be noted that the main idea of the following considerations is already published in Körner et al. [KBJS09].

4.4.1 Arrangement for the Distance

In order to distinguish between circles defined with respect to norm $\|\cdot\|$ and circles defined with respect to norm k we use the following notation.

Notation 4.11. Given $X \in \mathbb{R}^2$ and $r > 0$ we define

$$C(X, r) = \{Y : \|X - Y\| = r\},$$
$$C_k(X, r) = \{Y : k(X - Y) = r\}.$$

$C_k(X, r)$ is called k-circle. For $C(X, r)$ the denotations $\|\cdot\|$-*circle* and *circle* are used.

Note that $d_m(C(X, r))$ can be interpreted as the radius of the smallest k-circle with center A_m that touches the $\|\cdot\|$-circle $C(X, r)$; that is,

$$d_m(C(X, r)) = \min \left\{ v \geq 0 : C_{\|\cdot\|}(X, r) \cap C_k(A_m, v) \neq \emptyset \right\}.$$

Since $C_{\|\cdot\|} = C_{\|\cdot\|}(X, r)$ and $C_k = C_k(A_m, v)$ are polyhedrons which always intersect at a vertex of one of them, we can distinguish the following cases.

(i) C_k touches $C_{\|\cdot\|}$ in a vertex of $C_{\|\cdot\|}$. Let $E \in \mathrm{Ext}(B_1)$ be the fundamental direction of $\|\cdot\|$ that defines this vertex of $C_{\|\cdot\|}$. Then we have

$$d_m(C_{\|\cdot\|}(X, r)) = k(X + rE - A_m).$$

(ii) C_k touches $C_{\|\cdot\|}$ in a vertex of C_k. Let $F \in \text{Ext}(B_2)$ be the corresponding fundamental direction of k. Then we have

$$d_m(C_{\|\cdot\|}(X,r)) = \min\{|\lambda| : A_m + \lambda F \in C_{\|\cdot\|}\}.$$

For all $(X,r) \in \mathbb{R}^2 \times [0,\infty)$, $1 \leq m \leq M$, and $1 \leq g \leq s$ define the distance between the vertex $X + rE_g$ of the circle $C_{\|\cdot\|}(X,r)$ and the fixed point A_m in the following way

$$\gamma_{m,g}(X,r) := k(X + rE_g - A_m).$$

Furthermore, for all $(X,r) \in \mathbb{R}^2 \times [0,\infty)$, $1 \leq m \leq M$, and $1 \leq g \leq t$, define the distance between the circle $C_{\|\cdot\|}(X,r)$ and the fixed point A_m along the fundamental direction F_g in the following way

$$\delta_{m,g}(X,r) := \min\{|\lambda| : A_m + \lambda F_g \in C_{\|\cdot\|}(X,r)\}$$

where $\min \emptyset := \infty$.

With these notations we obtain the following result.

Lemma 4.12. *Let $\|\cdot\|$ and k be two polyhedral norms in $\mathbb{R}^2$. Furthermore, let $C_{\|\cdot\|}(X,r)$ be a circle and $A_m \in \mathscr{D}$ a fixed point. The point–circle distance between $C_{\|\cdot\|}(X,r)$ and A_m is given as*

$$d_m(C_{\|\cdot\|}(X,r)) = \min\left\{\min_{1\leq g\leq s} \gamma_{m,g}(X,r), \min_{1\leq g\leq t} \delta_{m,g}(X,r)\right\}.$$

Proof. It is clear that

$$\min_{1\leq g\leq s} \gamma_{m,g}(X,r) \geq d_m(C_{\|\cdot\|}(X,r)), \qquad \min_{1\leq g\leq t} \delta_{m,g}(X,r) \geq d_m(C_{\|\cdot\|}(X,r)).$$

As mentioned above, there exists a fundamental direction $E_g \in \text{Ext}(B_1)$ or $F_g \in \text{Ext}(B_2)$ such that either

$$\gamma_{m,g}(X,r) = d_m(C_{\|\cdot\|}(X,r)) \text{ or } \delta_{m,g}(X,r) = d_m(C_{\|\cdot\|}(X,r));$$

hence, we obtain the assertion. $\square$

For all $1 \leq m \leq M$, define

$$\delta_m(X,r) := \min_{1\leq g\leq s} \delta_{m,g}(X,r), \qquad \gamma_m(X,r) := \min_{1\leq g\leq t} \gamma_{m,g}(X,r).$$

We now identify an arrangement of $\mathbb{R}^2 \times [0,\infty[$ such that δ_m is concave on each 3-face and a second arrangement such that γ_m is concave on each 3-face. Then, combining both arrangements we obtain an arrangement of $\mathbb{R}^2 \times [0,\infty[$ such that $d_m(C(X,r)) = \min\{\delta_m(X,r), \gamma_m(X,r)\}$ is concave in (X,r) on each 3-face.

We start with the function δ_m. For all $1 \leq u \leq s$, $1 \leq v \leq t$, and $1 \leq m \leq M$, define

$$\mathcal{N}_m(u,v) := \left\{ \begin{pmatrix} A_m - \alpha_1 E_u + \alpha_2 F_v + \alpha_3 F_v^+ \\ \alpha_1 \end{pmatrix} \in \mathbb{R}^3 : \alpha_1, \alpha_2, \alpha_3 \geq 0 \right\}.$$

$\mathcal{N}_m(u,v)$ is a convex cone in $\mathbb{R}^3$ with apex $(A_m, 0)$. An example for $\mathcal{N}_m(u,v)$ is depicted in Fig. 4.2a. We show that $\gamma_{m,g}$ is affine linear on each cone $\mathcal{N}_m(g,v)$.

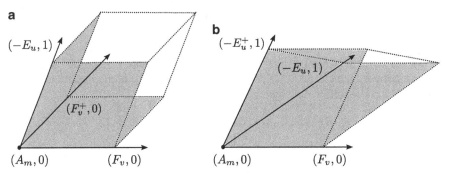

Fig. 4.2 Examples for cones of type $\mathcal{N}_m(u,v)$ and $\mathcal{M}_m(u,v)$. The *dashed lines* are reference lines to point out the dimensional shape of the cones. Facets of the boundaries of the cones are depicted in *dark gray*. (**a**) A cone of type $\mathcal{N}_m(u,v)$. (**b**) A cone of type $\mathcal{M}_m(u,v)$

Lemma 4.13. *Let $\| \cdot \|$ and k be two polyhedral norms in $\mathbb{R}^2$. Furthermore, let $m \in \{1, \ldots, M\}$ and $g \in \{1, \ldots, s\}$. Then for all $1 \leq v \leq t$, the function $\gamma_{m,g}(X,r)$ is affine linear in (X,r) on the cone $\mathcal{N}_m(g,v)$.*

Proof. Let $(X_1, r_1) \neq (X_2, r_2) \in \mathcal{N}_m(g,v)$ for some fixed values u, v. Then there exist nonnegative α_{k1}, α_{k2}, and α_{k3} such that

$$\begin{pmatrix} X_k \\ r_k \end{pmatrix} = \begin{pmatrix} A_m \\ 0 \end{pmatrix} + \alpha_{k1} \begin{pmatrix} -E_g \\ 1 \end{pmatrix} + \alpha_{k2} \begin{pmatrix} F_v \\ 0 \end{pmatrix} + \alpha_{k3} \begin{pmatrix} F_v^+ \\ 0 \end{pmatrix},$$

$k = 1, 2$. Since $\alpha_{k1} = r_k$ for $k = 1, 2$, we obtain

$$\gamma_{m,g}(X_1 + X_2 - A_m, r_1 + r_2)$$
$$= k(X_1 + X_2 + (r_1 + r_2)E_g - 2A_m)$$
$$= k((\alpha_{12} + \alpha_{22})F_v + (\alpha_{13} + \alpha_{23})F_v^+)$$
$$= \alpha_{12} + \alpha_{22} + \alpha_{13} + \alpha_{23}$$
$$= k(\alpha_{12}F_v + \alpha_{13}F_v^+) + k(\alpha_{22}F_v + \alpha_{23}F_v^+)$$
$$= k(X_1 + r_1 E_g - A_m) + k(X_2 + r_2 E_g - A_m)$$
$$= \gamma_{m,g}(X_1, r_1) + \gamma_{m,g}(X_2, r_2);$$

that is, $\gamma_{m,g}(X,r)$ is an affine linear function (with translation $(A_m,0)$) on each cone $\mathcal{N}_m(g,v)$, $1 \leq v \leq t$. $\qquad\square$

Now, consider the cones $\mathcal{N}_m(u,v)$, $1 \leq u \leq s$, $1 \leq v \leq t$. Each cone $\mathcal{N}_m(u,v)$ is spanned by three fundamental directions. Thus, three hyperplanes are induced by each cone $\mathcal{N}_m(u,v)$. Let $\mathcal{H}_{m1}$ be the set containing all hyperplanes induced by the cones $\mathcal{N}_m(u,v)$, $1 \leq u \leq s$, $1 \leq v \leq t$. The following result shows that $\mathcal{H}_{m1}$ induces an arrangement $\mathcal{A}^+(\mathcal{H}_{m1})$ such that $\gamma_m(X,r)$ is concave in X and r on any 3-face of $\mathcal{A}^+(\mathcal{H}_{m1})$.

Corollary 4.14. *Let $\|\cdot\|$ and k be two polyhedral norms in $\mathbb{R}^2$ and let $A_m \in \mathcal{D}$ be a fixed point. Then the function $\gamma_m(X,r)$ is concave in X and r on any 3-face of $\mathcal{A}^+(\mathcal{H}_{m1})$.*

Proof. Let $\mathcal{F}$ be a 3-face of $\mathcal{A}^+(\mathcal{H}_{m1})$. For all $g \in \{1,\dots,s\}$ there exists $v \in \{1,\dots,t\}$ such that

$$\mathcal{F} \subseteq \mathcal{N}_m(g,v).$$

Hence, it follows from Lemma 4.13 that $\gamma_{m,g}(X,r)$ is affine linear in X and r on $\mathcal{F}$. $\qquad\square$

Now, we consider the function $\delta_m(X,r)$. For all $1 \leq u \leq s$, $1 \leq v \leq t$, and $1 \leq m \leq M$, define

$$\delta_{m,v}^u(X,r) := \min\{\lambda \geq 0 : A_m + \lambda F_v \in S_u(X,r)\}$$

where

$$S_u(X,r) := \{X + \alpha_1 E_u + \alpha_2 E_u^+ : \alpha_1 + \alpha_2 = r, \alpha_1, \alpha_2 \geq 0\},$$

and $\min \emptyset := \infty$. Note that $S_u(X,r)$ is a facet of the circle $C_{\|\cdot\|}(X,r)$. Thus, $\delta_{m,v}^u(X,r)$ is the distance between the facet $S_u(X,r)$ of the circle $C_{\|\cdot\|}(X,r)$ and the fixed point A_m along the fundamental direction F_v. With this notation we obtain

$$\delta_m(X,r) = \min_{1 \leq u \leq s, \, 1 \leq v \leq t} \delta_{m,v}^u(X,r).$$

We derive an arrangement such that $\delta_{m,v}^u(X,r)$ is linear on any of its 3-faces. To this end, for all $1 \leq u \leq s$, $1 \leq v \leq t$, and $1 \leq m \leq M$, define

$$\mathcal{M}_m(u,v) := \left\{ \begin{pmatrix} A_m - \alpha_1 E_u - \alpha_2 E_u^+ + \alpha_3 F_v \\ \alpha_1 + \alpha_2 \end{pmatrix} : \alpha_1, \alpha_2, \alpha_3 \geq 0 \right\}.$$

Like the cone $\mathcal{N}_m(u,v)$ also $\mathcal{M}_m(u,v)$ is a convex cone in $\mathbb{R}^3$ with apex $(A_m,0)$, see Fig. 4.2b for an illustrative example. The following lemma shows that $\delta_{m,v}^u(X,r)$ is linear on the cone $\mathcal{M}_m(u,v)$.

Lemma 4.15. *Let $\|\cdot\|$ and k be two polyhedral norms in $\mathbb{R}^2$. Let $m \in \{1,\dots M\}$, and suppose that $u \in \{1,\dots,s\}$ and $v \in \{1,\dots,t\}$ such that $(F_v,0)$, $(-E_u,1)$, and $(-E_u^+,1)$ are linearly independent. Then the function $\delta_{m,v}^u(X,r)$ is linear on the cone $\mathcal{M}_m(u,v)$.*

Proof. Let u and v be fixed and assume that $(-E_u, 1), (-E_u^+, 1)$, and $(F_v, 0)$ are linearly independent. Let $(X, r) \in \mathcal{M}_m(u, v)$, i.e., there exist real values $\tilde{\alpha}_1, \tilde{\alpha}_2, \tilde{\alpha}_3 \geq 0$ such that

$$\binom{X}{r} = \binom{A_m}{0} + \tilde{\alpha}_1 \binom{-E_u}{1} + \tilde{\alpha}_2 \binom{-E_u^+}{1} + \tilde{\alpha}_3 \binom{F_v}{0}. \tag{4.2}$$

We show $\delta_{m,v}^u(X, r) = \tilde{\alpha}_3$ for all $(X, r) \in \mathcal{M}_m(u, v)$. To this end note that $A_m + \lambda F_v \in S_u(X, r)$ if and only if

$$X - \lambda F_v \in \overline{S}_u(A_m, r) = \{A_m - \alpha_1 E_u - \alpha_2 E_u^+ : \alpha_1 + \alpha_2 = r, \alpha_1, \alpha_2 \geq 0\}.$$

Therefore, we obtain

$$\delta_{m,v}^u(X, r) = \min\{\lambda \geq 0 : X - \lambda F_v \in \overline{S}_u(A_m, r)\}.$$

Since $A_m - \tilde{\alpha}_1 E_u - \tilde{\alpha}_2 E_u^+ \in \overline{S}_u(A_m, r)$ we conclude $X - \tilde{\alpha}_3 F_v \in \overline{S}_u(A_m, r)$ and hence $\delta_{m,v}^u(X, r) \leq \tilde{\alpha}_3$. Due to the linear independence of

$$(-E_u, 1), (-E_u^+, 1), (F_v, 0)$$

the representation of (X, r) given in (4.2) is unique. Therefore we may conclude $\delta_{m,v}^u(X, r) = \tilde{\alpha}_3$ and hence we obtain the assertion. $\qquad\square$

Now, we define an arrangement $\mathcal{A}^+(\mathcal{H}_{m2})$ analogously to the arrangement $\mathcal{A}^+(\mathcal{H}_{m1})$; that is, we define $\mathcal{H}_{m2}$ as the set of hyperplanes induced by the cones $\mathcal{M}_m(u, v)$, $1 \leq u \leq s$, $1 \leq v \leq t$. Let $\mathcal{F} \in \mathcal{A}^+(\mathcal{H}_{m2})$ be a 3-face and define the set

$$\mathcal{J}_1 := \{(u, v) \in \{1, \ldots, s\} \times \{1, \ldots, t\} : \mathcal{F} \cap \mathrm{int}(\mathcal{M}_m(u, v)) = \emptyset\}.$$

Furthermore, define $\mathcal{J}_2 := \{1, \ldots, s\} \times \{1, \ldots, t\} \setminus \mathcal{J}_1$. Then we obtain

$$\delta_{m,v}^u(X, r) = \infty \text{ for all } \binom{X}{r} \in \mathcal{F} \quad \Longleftrightarrow \quad (u, v) \in \mathcal{J}_1.$$

Using the sets $\mathcal{J}_1$ and $\mathcal{J}_2$ we can prove the following lemma.

Lemma 4.16. *Let $\|\cdot\|$ and k be two polyhedral norms in $\mathbb{R}^2$ and let $A_m \in \mathcal{D}$ be a fixed point. Then $\delta_m(X, r)$ is a concave function in X and r on any 3-face of the arrangement $\mathcal{A}^+(\mathcal{H}_{m2})$.*

Proof. Let $\mathcal{F}$ be a 3-face of the arrangement $\mathcal{A}^+(\mathcal{H}_{m2})$ and let $(X, r) \in \mathrm{int}(\mathcal{F})$. Consider the circle $C(A_m, r)$, the point X, and the direction lines of norm k through X. As depicted in Fig. 4.3 two cases are possible:

- There exists a fundamental direction F_v of norm k such that the ray $[X, X - F_v\rangle$ intersects $C_{\|\cdot\|}(A_m, r)$. This situation applies for the point X_1 in Fig. 4.3.
- For all fundamental directions F_v, $1 \leq v \leq t$ of the polyhedral norm k we have $[X, X - F_v\rangle \cap C_{\|\cdot\|}(A_m, r) = \emptyset$. This situation applies for the point X_2 in Fig. 4.3.

In the former case we conclude $\mathscr{J}_2 \neq \emptyset$ and obtain

$$\delta_m(X,r) = \min_{(u,v)\in\mathscr{J}_2} \delta^u_{m,v}.$$

For $(X,r) \in \text{int}(\mathscr{F})$ the vectors $(-E_u,1)$, $(-E_u^+,1)$, and $(F_v,0)$ are linearly independent for all $(u,v) \in J_2$. Hence for all $(u,v) \in J_2$ the function $\delta^u_{m,v}$ is affine linear. Therefore, we obtain that $\delta_m(X,r)$ is concave on $\mathscr{F}$.

In the latter case we may conclude $\mathscr{J}_2 = \emptyset$. Hence, $\delta_m(X,r) = \infty$ for all points $(X,r) \in \mathscr{F}$.

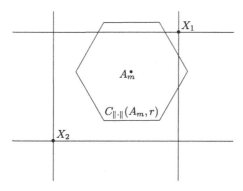

Fig. 4.3 Direction lines of norm $k = \ell_1$ through X_1 and X_2 and a circle $C_{\|\cdot\|}(A_m,r)$

Summarizing, on any 3-face of the arrangement $\mathscr{A}^+(\mathscr{H}_{m2})$, $\delta_m(X,r)$ is concave. Note that it is possible that $\delta_m(X,r)$ has everywhere on $\mathscr{F}$ the value $+\infty$. □

Using Lemma 4.14 and Lemma 4.16 we obtain the following result.

Theorem 4.17. *Let $\|\cdot\|$ and k be two polyhedral norms in $\mathbb{R}^2$. Furthermore, let $A_m \in \mathscr{D}$ be a fixed point and let $\mathscr{A}^+(\mathscr{H})$ be the arrangement of $\mathbb{R}^2 \times [0,\infty[$ we obtain by combining $\mathscr{A}^+(\mathscr{H}_{m1})$ and $\mathscr{A}^+(\mathscr{H}_{m2})$. Then the point–circle distance $d_m(C(X,r))$ is concave on each 3-face $\mathscr{F} \in \mathscr{A}^+(\mathscr{H})$.*

Remark 4.18. From Lemma 4.3 and Theorem 4.17 we obtain that $d_m(C(X,r))$ is affine linear on each 3-face $\mathscr{F}$ of the arrangement $\mathscr{A}^+(\mathscr{H})$, provided that $(A_m,0) \notin \text{int}(\mathscr{F})$.

4.4.2 A Finite Dominating Set

With Theorem 4.17 it is straightforward to obtain an arrangement for the minisum circle problem with unequal polyhedral norms. We just define $\mathscr{A}^+(\mathscr{H})$ to be the arrangement we obtain by combining the arrangements $\mathscr{A}(\mathscr{H}_m)$ for $1 \leq m \leq M$.

Then the point–circle distance $d_m(C(X,r))$ is concave for all $1 \leq m \leq M$ on any 3-face $\mathscr{F} \in \mathscr{A}(\mathscr{H})$. Thus, also the objective function

$$f(C(X,r)) = \sum_{i=1}^{n} \omega_m d_m(C(X,r))$$

of the minisum circle problem with unequal polyhedral norms is concave on any 3-face $\mathscr{F} \in \mathscr{A}(\mathscr{H})$. Therefore we obtain the following results.

Theorem 4.19. *There exists a minisum circle for the minisum circle problem with unequal polyhedral norms.*

Proof. The objective function is bounded below by zero and concave on each 3-face of the arrangement $\mathscr{A}(\mathscr{H})$. □

Theorem 4.20. *Let $\|\cdot\|$ and k be two polyhedral norms in $\mathbb{R}^2$. Furthermore, let $\mathscr{V}$ denote the set that contains all vertices of the arrangement $\mathscr{A}(\mathscr{H})$. Then the function Ξ which maps each instance of the minisum circle problem with unequal polyhedral norms onto the corresponding set $\mathscr{V}$ is a finite dominating set for the minisum circle problem with unequal norms.*

Lemma 4.21. *For each instance of the minisum circle problem with unequal polyhedral norms the set $\mathscr{V}$ contains at most $\mathcal{O}((Mst)^3)$ points.*

Proof. Each cone $\mathscr{N}_m(u,v)$, $\mathscr{M}_m(u,v)$ induces exactly three hyperplanes. Any point in $\mathscr{V}$ is the intersection of three different hyperplanes. There are at most $\mathcal{O}(Mst)$ different hyperplanes. Hence, we have at most $\mathcal{O}((Mst)^3)$ points in $\mathscr{V}$. □

The structure of the arrangement $\mathscr{A}^+(\mathscr{H})$ gives rise to the following result.

Lemma 4.22. *There exists a minisum circle $C(X,r)$ for the minisum circle problem with unequal polyhedral norms $\|\cdot\|$ and k such that at least μ fixed points A_{m_i} are contained in $C(X,r)$, i.e., $A_{m_i} \in C(X,r)$, $1 \leq i \leq \mu$. We have $0 \leq \mu \leq 3$ and at least $3 - \mu$ vertices of $C(X,r)$ lie on the two-dimensional grid formed by all direction lines of norm k through the fixed points A_m, $1 \leq m \leq M$.*

Proof. Due to Theorem 4.20 there exists a minisum circle $C(X,r)$ such that $(X,r) \in \mathscr{V}$. As mentioned before, any point $(X,r) \in \mathscr{V}$ is the intersection of (at least) three different hyperplanes induced by some cones of the type $\mathscr{N}_m(u,v)$ and $\mathscr{M}_m(u,v)$, say $H_{m_1}, H_{m_2}, H_{m_3}$. For the sake of simplicity we assume that (X,r) is the intersection of exactly three hyperspheres; that is, no other hyperplane than H_{m_1}, H_{m_2}, and H_{m_3} contains (X,r) (otherwise we may remove this hyperplane from $\mathscr{H}$).

Assume that hyperplane H_{m_i} is induced by the cone $\mathscr{N}_m(u,v)$. Since $r > 0$ we obtain that there exists $\lambda \in \mathbb{R}$ such that

$$\begin{pmatrix} A_{m_i} \\ 0 \end{pmatrix} + r \begin{pmatrix} E_u \\ 1 \end{pmatrix} + \lambda \begin{pmatrix} F_v \\ 0 \end{pmatrix} = \begin{pmatrix} X \\ r \end{pmatrix}, \tag{4.3}$$

i.e., the vertex $X + rE_u$ of $C(X,r)$ lies on the direction line $\langle A_{m_i}, A + F_v \rangle$.

If hyperplane H_{m_i} is induced by the cone $\mathcal{M}_m(u,v)$, then two cases are possible. Either there exists $\lambda \in \mathbb{R}$ such that (X,r) has representation (4.3), or there exist $\lambda_u, \lambda_v \in \mathbb{R}$ such that

$$\begin{pmatrix} A_{m_i} \\ 0 \end{pmatrix} + \lambda_u \begin{pmatrix} E_u \\ 1 \end{pmatrix} + \lambda_v \begin{pmatrix} E_u^+ \\ 1 \end{pmatrix} = \begin{pmatrix} X \\ r \end{pmatrix}.$$

In the former case we obtain that $X + rE_u$ lies on the direction line $\langle A_{m_i}, A + F_v \rangle$. In the latter case we may assume $\lambda_u, \lambda_v \geq 0$. Indeed, $\lambda_u < 0$ or $\lambda_v < 0$ implies that (X,r) does not belong to the cone $\mathcal{M}_m(u,v)$. Hence, we can remove H_{m_i} from $\mathcal{H}$ and obtain that $f(C(X,r))$ is concave in a region around (X,r), i.e., $C(X,r)$ is either no minisum circle or there exists another vertex of the arrangement $\mathcal{A}^+(\mathcal{H})$ where f attains a global minimum. $\lambda_u, \lambda_v \geq 0$ imply that A_{m_i} is contained in $C(X,r)$. $\quad\square$

4.5 Concluding Remarks

The generalization of the minisum circle problem to the model discussed in this chapter leads to a significantly more difficult problem. The point–circle distance cannot be evaluated as easily as it was the case for the original model. However, some results of the minisum circle problem extend to the problem with unequal norms. The distance is still symmetric and has the same convexity and concavity properties. A point cannot be a degenerated optimal solution and the dominance criterion for the minisum circle problem also extends to the problem with unequal norms. For the case of two unequal polyhedral norms a finite dominating set was derived. This result gives rise to a simple $\mathcal{O}(M^4)$ brute-force algorithm. Other solution approaches are possible for special cases as for instance the Tschebyschow–Manhattan (Manhattan–Tschebyschow) case where norm $\| \cdot \|$ is the Tschebyschow (Manhattan) norm and norm k is the Manhattan (Tschebyschow) norm. We discuss the Tschebyschow–Manhattan case in the following chapter.

Possible lines of further research include the generalization of the problem to higher dimensions $n \geq 3$. In order to approximate an optimal solution, methods of global optimization seem to be promising.

A natural extension of the problem is to replace the norms $\| \cdot \|$ and k by convex distance functions. As a first step, only the norm $\| \cdot \|$ should be replaced. In particular for distance functions defined with respect to convex polytopes, i.e., for polyhedral gauges, most results of this chapter should remain valid.

Chapter 5
Minisum Rectangles in a Manhattan Plane

5.1 Basic Assumptions

In this chapter we consider the problem of locating a rectangle. Let $\mathscr{R}$ denote the set of nondegenerated, axis-parallel rectangles in $\mathbb{R}^2$. Then the setting of this chapter can be described in the following way:

Dimensions: $n = 2, M \geq 2$
Objects: Subset of nondegenerated, axis-parallel rectangles $\mathscr{R}$
Metric: Induced by the Manhattan norm $\|\cdot\|_1$; $d(X,Y) := \|Y - X\|_1$
Distance: $d(R,A_m) = \min\{d(Y,A_m) : Y \in R\}, 1 \leq m \leq M$
Objective: $f(R) = \sum_{m=1}^{M} \omega_m d(R,A_m) \rightarrow \min$

In contrast to previous chapters we do not use a norm $\|\cdot\|$ to define hyperspheres as the objects we want to locate, but we consider subsets of the set of axis-parallel rectangles $\mathscr{R}$ in $\mathbb{R}^2$. Given a subset $\mathscr{R}' \subseteq \mathscr{R}$ we search an optimal rectangle among all rectangles in $\mathscr{R}'$ which minimizes the minisum criterion. Nevertheless, both approaches are quite related. In particular, it will turn out that the present problem is equivalent to the minisum circle problem with unequal norms for a certain subset of $\mathscr{R}$. In order to see the relation between the present problem and the theory of previous chapters, we define *weighted Tschebyschow norms*. A weighted Tschebyschow norm is a Tschebyschow norm with additional weight for each fundamental direction. Formally, we define a weighted Tschebyschow norm in the following way.

Definition 5.1. Let $C = (c_1, \ldots, c_n) \in \mathbb{R}^n$ such that $c_i > 0$ for all $1 \leq i \leq n$. For all $X = (x_1, \ldots, x_n) \in \mathbb{R}^n$ let

$$\|X\|_\infty^C = \max\{c_i|x_i| : 1 \leq i \leq n\}.$$

$\|X\|_\infty^C$ is a norm in $\mathbb{R}^n$ which is called *weighted Tschebyschow norm*.

M.-C. Körner, *Minisum Hyperspheres*, Springer Optimization and Its Applications 51, 77
DOI 10.1007/978-1-4419-9807-1_5, © Springer Science+Business Media, LLC 2011

Note that a circle defined with respect to a weighted Tschebyschow norm is a closed, convex quadrilateral with four right angles, i.e., it is an axis-parallel rectangle. Therefore, the set of axis-parallel rectangles $\mathscr{R}$ can be described by weighted Tschebyschow norms. More precisely, the set $\mathscr{R}$ contains all scaled and translated unit circles of weighted Tschebyschow norms in $\mathbb{R}^2$. Due to close relation between axis-parallel rectangles in $\mathbb{R}^2$ and weighted Tschebyschow norms we can use the results of the previous chapter in order to derive results for the model discussed in this chapter. In the following we consider three restricted problems where feasible rectangles $\mathscr{R}'$ are subsets of $\mathscr{R}$:

1. Feasible rectangles are *axis-parallel* and have *equal perimeter* and *equal aspect ratio*. We denote the set of these rectangles as $\mathscr{R}_{a,p}$ where $p > 0$ is the perimeter and $a > 0$ is the aspect ratio.
2. Feasible rectangles are *axis-parallel* and have *equal aspect ratio*. This set is denoted as $\mathscr{R}_a$ where $a > 0$ is the aspect ratio.
3. Feasible rectangles are *axis-parallel* and have *equal perimeter*. For this set of rectangles we use the denotation $\mathscr{R}_p$ where $p > 0$ is the perimeter of the rectangles in $\mathscr{R}_p$.

In addition, we also consider the unrestricted problem where any rectangle $R \in \mathscr{R}$ is a feasible solution.

For all $\mathscr{R}' \in \{\mathscr{R}_{a,p}, \mathscr{R}_a, \mathscr{R}_p, \mathscr{R}\}$, we denote the problem

$$\min_{R \in \mathscr{R}'} f(R) = \sum_{m=1}^{M} \omega_m d(R, A_m)$$

as *rectangle location problem with feasible set $\mathscr{R}'$*. An optimal solution to this problem is denoted as *minisum rectangle*. Rectangle location problems have applications for example in connection with paper position sensing. Here, a rectangle location problem is used in order to determine the position of a sheet of paper in a paper moving system, see [BG02]. Other applications include control software for a sawyer motor robot [WCM93] and, more general, the recognition of geometric objects [Gri91].

In the following we study properties of minisum rectangles for the feasible sets mentioned above. Initially, we introduce some notations and consider the *point–rectangle distance*.

5.2 Notations

We use the letter R in order to refer to the elements of $\mathscr{R}$. For the sake of simplification and clarification we introduce the following notation.

Notation 5.2. For any $c_1, c_2 > 0$, let $\mathscr{G}_{c_1,c_2}$ denote the set which consists of all scaled and translated unit circles of the norm $\| \cdot \|_{\infty}^{c_1,c_2}$.

As mentioned above, there exists a strong relation between the set $\mathcal{R}$ and sets of type $\mathcal{G}_{c_1,c_2}$. Indeed, we have

$$\mathcal{R} = \bigcup_{c_1,c_2 > 0} \mathcal{G}_{c_1,c_2};$$

that is, any rectangle $R \in \mathcal{R}$ has a representation of the form

$$R = \lambda C_0 + \{X\} \tag{5.1}$$

where $\lambda > 0$, $X \in \mathbb{R}^2$, and C_0 is the unit circle of a weighted Tschebyschow norm. That means that any rectangle can be interpreted as a circle defined with respect to a weighted Tschebyschow norm. In order to indicate that circles defined with respect to weighted Tschebyschow norms are rectangles we do not use the notation of previous chapters. Instead, we introduce the following notation.

Notation 5.3. Given $c_1, c_2, r > 0$ and $X \in \mathbb{R}^2$ let $R_{c_1,c_2}(X,r)$ denote the rectangle which is given by (5.1); that is,

$$R_{c_1,c_2}(X,r) = \{Y \in \mathbb{R}^2 : \|Y - X\|_\infty^{c_1,c_2} = r\}.$$

Also the notation $R(X,r)$ is used in order to refer to $R_{c_1,c_2}(X,r)$, provided that this notation is nonambiguous. X is denoted as *center* and r as *radius* of the rectangle R.

Any rectangle in $R \in \mathcal{R}$ may be described according to Notation 5.3. Note that such a representation is not unique; that is, there exists an unbounded number of pairs (c_1,c_2) such that $R = R_{c_1,c_2}(X,r)$ for a rectangle $R \in \mathcal{R}$. A rectangle $R \in \mathcal{R}$ has a unique representation if we set $r = 1$, i.e., there exists a unique pair (c_1,c_2) such that $R = R_{c_1,c_2}(X,1)$ for some $X \in \mathbb{R}^2$.

In order to refer to an appropriate weighted Tschebyschow norm for the representation of all rectangles in a set of rectangles we introduce the following definition.

Definition 5.4. Let $\mathcal{R}' \subseteq \mathcal{R}$ be a set of nondegenerated, axis-parallel rectangles. If there exist $c_1, c_2 > 0$ such that $R \in \mathcal{G}_{c_1,c_2}$ for all $R \in \mathcal{R}'$ then we denote the corresponding weighted Tschebyschow norm $\|\cdot\|_\infty^{c_1,c_2}$ as $\mathcal{R}'$-*generating norm*.

On the one hand, for a set $\mathcal{R}' \subseteq \mathcal{R}$ there may exist an unbounded number of $\mathcal{R}'$-generating norms. On the other hand, it is possible that there does not exist any $\mathcal{R}'$-generating norm for a set $\mathcal{R}' \subseteq \mathcal{R}$. We will see examples for both cases in the following.

Given a rectangle $R \in \mathcal{R}$ and a weighted Tschebyschow norm with weights c_1, c_2 the perimeter p of R is given as

$$p = \frac{2}{c_1} + \frac{2}{c_2}.$$

The aspect ratio of R may be defined in several ways. We use the following definition.

Definition 5.5. Let $R \in \mathcal{R}$ be an axis-parallel rectangle and let c_1, c_2 be the weights of a weighted Tschebyschow norm $\|\cdot\|_\infty^{c_1,c_2}$ such that $R = R_{c_1,c_2}(X,r)$ for some $X \in \mathbb{R}^2$, $r > 0$. Then we define the *aspect ratio* a of R by $a = c_2/c_1$.

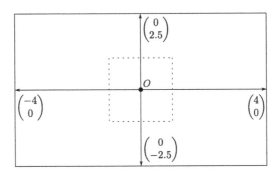

Fig. 5.1 The *dashed rectangle* represents the unit rectangle of the Tschebyschow norm. The unit rectangle $R_{c_1,c_2}(O,1)$ of a weighted Tschebyschow norm with weights $c_1 = \frac{1}{4}$ and $c_2 = \frac{2}{5}$ is depicted *solid*. The side lengths of this rectangle are given by $2c_1^{-1} = 8$ and $2c_2^{-1} = 5$, respectively. The aspect ratio is $8/5$

According to Definition 5.5 the rectangle depicted in Fig. 5.1 has aspect ratio $a = 8/5$. Note that this ratio equals the ratio between the side length in x-direction and the side length in y-direction.

5.3 Point–Rectangle Distance

We define the distance between a rectangle and a fixed point just like the distance between a circle and a point; that is, the distance between a rectangle $R \in \mathcal{R}$ and a fixed point $A_m \in \mathcal{D}$ is given by

$$d_m(R) = d(R,A_m) = \min\{\|Y - A_m\|_1 : Y \in R\}$$

where $\|\cdot\|_1$ denotes the Manhattan norm. Due to the fact that $\|X\|_1 = |x_1| + |x_2|$ for any $X = (x_1,x_2) \in \mathbb{R}^2$ the distance between a rectangle $R \in \mathcal{R}$ and a point $A = (a_1,a_2)$ may also be written as

$$d(R,A) = \min\{|y_1 - a_1| + |y_2 - a_2| : (y_1,y_2) \in R\}.$$

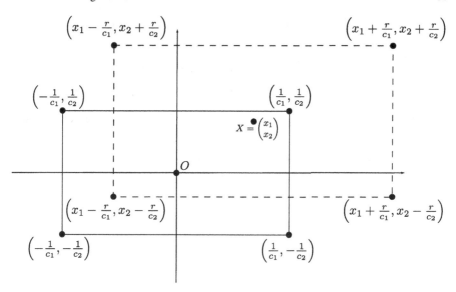

Fig. 5.2 Relation between the weights c_1, c_2 of a weighted Tschebyschow norm $\| \cdot \|_\infty^{c_1, c_2}$ and the vertices of the unit rectangle $R_{c_1, c_2}(O, 1)$ and the rectangle $R_{c_1, c_2}(X, r)$, respectively

Figure 5.2 shows the relation between the weights c_1, c_2 of $\| \cdot \|_\infty^{c_1, c_2}$ and the vertices of the unit rectangle $R_{c_1, c_2}(O, 1)$ and the vertices of an arbitrary rectangle $R_{c_1, c_2}(X, r)$. We use these relations in order to derive a simple formula for the distance between a point $A_m \in \mathscr{D}$ and a rectangle $R \in \mathscr{R}$. To this end let $R \in \mathscr{R}$ be fixed and suppose that $R = R_{c_1, c_2}(X, r)$ for some $X \in \mathbb{R}^2$, $r, c_1, c_2 > 0$. Furthermore, let x_1, x_2 denote the components of $X \in \mathbb{R}^2$ and define a partition of the plane into the four regions $\mathscr{U}_1(X, r)$, $\mathscr{U}_2(X, r)$, $\mathscr{U}_3(X, r)$, and $\mathscr{U}_4(X, r)$ depicted in Fig. 5.3. Formally, these regions are given by the following equations:

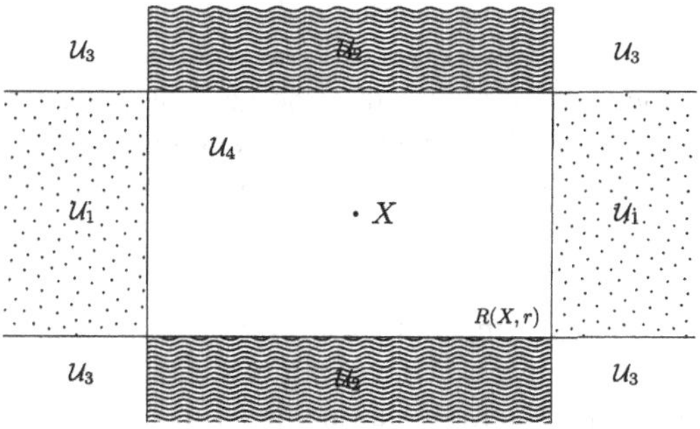

Fig. 5.3 Partition of the plane into the regions $\mathscr{U}_1 = \mathscr{U}_1(X, r)$, $\mathscr{U}_2 = \mathscr{U}_2(X, r)$, $\mathscr{U}_3 = \mathscr{U}_3(X, r)$, and $\mathscr{U}_4 = \mathscr{U}_4(X, r)$

$$\mathcal{U}_1(X,r) = \{(y_1,y_2) \in \mathbb{R}^2 \ : \ c_1|y_1 - x_1| \geq r, \ c_2|y_2 - x_2| \leq r\},$$
$$\mathcal{U}_2(X,r) = \{(y_1,y_2) \in \mathbb{R}^2 \ : \ c_1|y_1 - x_1| \leq r, \ c_2|y_2 - x_2| \geq r\},$$
$$\mathcal{U}_3(X,r) = \{(y_1,y_2) \in \mathbb{R}^2 \ : \ c_1|y_1 - x_1| \geq r, \ c_2|y_2 - x_2| \geq r\},$$
$$\mathcal{U}_4(X,r) = \{(y_1,y_2) \in \mathbb{R}^2 \ : \ c_1|y_1 - x_1| \leq r, \ c_2|y_2 - x_2| \leq r\}.$$

Using this partition and elementary properties of the Manhattan norm we obtain the following result.

Lemma 5.6. *Let $A = (a_1, a_2) \in \mathbb{R}^2$ be a point. Then we have the following formula for the distance between A and the rectangle $R = R_{c_1,c_2}(X,r)$:*

$$d(R,A) = \begin{cases} |a_1 - x_1| - \frac{r}{c_1} & \text{if } A \in \mathcal{U}_1(X,r) \\ |a_2 - x_2| - \frac{r}{c_2} & \text{if } A \in \mathcal{U}_2(X,r) \\ |a_1 - x_1| - \frac{r}{c_1} + |a_2 - x_2| - \frac{r}{c_2} & \text{if } A \in \mathcal{U}_3(X,r) \\ \min\left\{\frac{r}{c_1} - |a_1 - x_1|, \ \frac{r}{c_2} - |a_2 - x_2|\right\} & \text{if } A \in \mathcal{U}_4(X,r) \end{cases}.$$

Proof. Let $Y \in R = R_{c_1,c_2}(X,r)$ such that $d(R,A) = \|Y - A\|_1$.

If $A \in \mathcal{U}_1(X,r)$ then Y lies on a vertical edge $\mathscr{E}_v$ of R. Hence, $\|Y - A\|_1$ is the horizontal distance between A and R, i.e., $\|Y - A\|_1 = |a_1 - y_1|$ where $A = (a_1, a_2)$ and $Y = (y_1, y_2)$. Furthermore, $Y \in \mathscr{E}_v$ implies $|x_1 - y_1| = r/c_1$ where $X = (x_1, x_2)$. We distinguish the case $a_1 < x_1$ and the case $a_1 > x_1$. In the former case we have $a_1 \leq y_1 < x_1$ and obtain

$$\|Y - A_1\|_1 = y_1 - a_1 = -a_1 + x_1 - \frac{r}{c_1} = |a_1 - x_1| - \frac{r}{c_1}.$$

In the latter case we have $a_1 \geq y_1 > x_1$ and obtain $\|Y - A_1\| = a_1 - x_1 - r/c_1$. Thus, we have $d(R,A) = |a_1 - x_1| - r/c_1$.

If $A \in \mathcal{U}_2(X,r)$ then Y lies on a horizontal edge of R. This case can be handled analogously to the previous case.

If $A \in \mathcal{U}_3(X,r)$ then Y has to be a vertex of R, i.e., $y_i = x_i \pm r/c_i$, $i = 1,2$. Hence, the formula $d(R,A) = |a_1 - x_1| - r/c_1 + |a_2 - x_2| - r/c_2$ can be readily verified.

If $A \in \mathcal{U}_4(X,r)$ then $\|Y - A\|_1$ is either the vertical distance or the horizontal distance between R and A. In the former case we obtain $d(R,A) = r/c_1 - |a_1 - x_1|$. In the latter case we obtain $d(R,A) = r/c_2 - |a_2 - x_2|$. Hence, we conclude $d(R,A) = \min\left\{\frac{r}{c_1} - |a_1 - x_1|, \ \frac{r}{c_2} - |a_2 - x_2|\right\}$. $\qquad\square$

As it can be read from Fig. 5.3 the sets $\mathcal{U}_i(X,r)$, $i = 1,2$, consist of two and the set $\mathcal{U}_3(X,r)$ consists of four connected convex regions. From Lemma 5.6 we may conclude that the distance $d(R(X,r),A)$ is affine linear in A on each connected region of $\mathcal{U}_i(X,r)$, $i = 1,2,3$. Furthermore, it can be obtained from Lemma 5.6 (and also from Lemma 4.3) that $d(R(X,r),A)$ is concave in A on $\mathcal{U}_4(X,r)$. Hence, we have the following results.

Corollary 5.7. *The point–rectangle distance $d(R(X,r),A)$ is affine linear in A on any connected region of $\mathcal{U}_i(X,r)$, $i = 1,2,3$, and it is concave in A on $\mathcal{U}_4(X,r)$.*

Corollary 5.8. *The point–rectangle distance $d(R(X,r),A)$ is affine linear in X on any connected region of $\mathcal{U}_i(A,r)$, $i = 1,2,3$, and it is concave in X on $\mathcal{U}_4(A,r)$.*

Proof. Due to the fact that the point–rectangle distance $d(R(X,r),A)$ is symmetrical in X and A (cf. Lemma 4.1) the assertion follows from Lemma 5.6. □

Note that Lemma 4.3 implies that for fixed A the distance $d(R(X,r),A)$ is concave in X and r on the set

$$\mathcal{V} = \{(Y,s) \in \mathbb{R}^2 \times [0,\infty[\ : \ \|Y - A\| \le s\} \subseteq \mathbb{R}^3 \tag{5.2}$$

and convex in X and r on any convex set $U \subseteq \mathbb{R}^2 \times [0,\infty[\setminus \mathcal{V}$. This result can be strengthened. To this end, let $\mathscr{A}^+(\mathscr{H}_m)$ denote the arrangement defined in Section 4.4; that is, $\mathscr{A}^+(\mathscr{H}_m)$ is induced by hyperplanes containing two-dimensional facets of cones $\mathscr{N}_m(u,v)$, $1 \le u < v \le 4$. Recall that we refer to an l-dimensional element of $\mathscr{A}^+(\mathscr{H}_m)$ as l-face, $0 \le l \le 3$, and call a 0-face vertex. Further, recall that $\mathscr{A}^+(\mathscr{H}_m)$ is an arrangement of $\mathbb{R}^2 \times [0,\infty[$. The construction of $\mathscr{A}^+(\mathscr{H}_m)$ allows that any point (X,r) belonging to an l-face, $0 \le l \le 3$, of $\mathscr{A}^+(\mathscr{H}_m)$ can be identified with a rectangle $R_{c_1,c_2}(X,r) \in \mathscr{G}_{c_1,c_2}$. Note that the arrangement $\mathscr{A}^+(\mathscr{H}_m)$ depends on the norm $\|\cdot\|_\infty^{c_1,c_2}$; that is, $\mathscr{A}^+(\mathscr{H}_m)$ changes if we consider different weighted Tschebyschow norms.

Lemma 5.9. *Let $A_m \in \mathscr{D}$ be a fixed point and define $\mathcal{V}$ according to (5.2). Furthermore, let $\mathscr{F}$ be any 3-face of $\mathscr{A}^+(\mathscr{H}_m)$ such that $\operatorname{int}(\mathcal{V}) \cap \mathscr{F} = \emptyset$. Then the point–rectangle distance $d(R(X,r),A_m)$ is affine linear in X and r on $\mathscr{F}$.*

Proof. Let $\mathcal{U}_i^j(A_m,r)$, $j = 1,\dots,4$, denote the greatest connected open region of $\mathcal{U}_i(A_m,r)$, $i = 1,2,3$, where $\mathcal{U}_i^3(A_m,r) = \mathcal{U}_i^4(A_m,r) = \emptyset$, provided that $i \in \{1,2\}$. With $\|\cdot\| = \|\cdot\|_\infty^{c_1,c_2}$ and $k = \|\cdot\|_1$ it can be observed that each 3-face $\mathscr{F}$ of the arrangement $\mathscr{A}^+(\mathscr{H}_m)$ is a generalization of the regions $\mathcal{U}_i(A_m,r)$, $1 \le i \le 4$, in the sense that $\mathscr{F}$ has a representation either of the form

$$\mathscr{F} = \bigcup_{r \ge 0} \left\{ (Y,s) \in \mathbb{R}^2 \times [0,\infty[\ : \ Y \in \mathcal{U}_i^j(A_m,s) \right\} \tag{5.3}$$

for fixed $i = 1,2,3$ and fixed $j = 1,\dots,4$, or

$$\mathscr{F} = \mathcal{V} = \bigcup_{r \ge 0} \{ (Y,s) \in \mathbb{R}^2 \times [0,\infty[\ : \ Y \in \mathcal{U}_4(A_m,s) \}. \tag{5.4}$$

The same observation can be made for the arrangement which is induced by hyperplanes containing two-dimensional facets of cones $\mathscr{M}_m(u,v)$, $1 \le u < v \le 4$ (cf. Section 4.4). Hence, we can apply Theorem 4.17 and obtain that $d_m(R(X,r))$ is concave in X and r on each 3-face of the arrangement $\mathscr{A}^+(\mathscr{H}_m)$. Together with the convexity result of Lemma 4.3 we obtain that $d_m(R(X,r))$ is affine linear in X and r on $\mathscr{F}$, provided that $\mathscr{F} \cap \operatorname{int}(\mathcal{V}) = \emptyset$. □

Recall that the same result also applies to the point–rectangle distance considered in Chapter 4, see Remark 4.18. Also note that $\mathscr{F} \cap \mathscr{V} \neq \emptyset$ implies that $(A_m, 0) \in \mathscr{F}$.

5.4 Restricted Problems

We now discuss the three restricted rectangle location problems with feasible sets $\mathscr{R}_{a,p}$, $\mathscr{R}_a$, and $\mathscr{R}_p$.

5.4.1 Equal Aspect Ratio and Perimeter

We start with the rectangle location problem with feasible set $\mathscr{R}_{a,p}$ which consists of rectangles with aspect ratio $a > 0$ and perimeter $p > 0$. The following result shows that for any set $\mathscr{R}_{a,p}$ there exists a unique weighted Tschebyschow norm $\|\cdot\|_\infty^{c_1,c_2}$ such that $R = R_{c_1,c_2}(X,1)$ for all $R \in \mathscr{R}_{a,p}$; that is, $\|\cdot\|_\infty^{c_1,c_2}$ is a unique $\mathscr{R}_{a,p}$-generating norm with this property.

Lemma 5.10. *For any $a, p > 0$, there exists a unique pair $c_1, c_2 > 0$ such that*

$$\mathscr{R}_{a,p} = \{R_{c_1,c_2}(X,1) : X \in \mathbb{R}^2\}. \tag{5.5}$$

Furthermore, the weighted Tschebyschow norm $\|\cdot\|_\infty^{c_1,c_2}$ is an $\mathscr{R}_{a,p}$-generating norm.

Proof. Let R_0 be the unique rectangle with aspect ratio a and perimeter p having its center at the origin $O \in \mathbb{R}^2$. Define

$$c_1 = \frac{2 + 2a^{-1}}{p}, \quad c_2 = c_1 a.$$

Then we have $R_0 = R_{c_1,c_2}(O,1)$; that is, $\|\cdot\|_\infty^{c_1,c_2}$ is an $\{R_0\}$-generating norm. Since R_0 is unique also $\|\cdot\|_\infty^{c_1,c_2}$ is unique. Due to the fact that any rectangle $R \in \mathscr{R}_{a,p}$ can be represented as translation of $R_0 = R_{c_1,c_2}(O,1)$ we conclude that $\|\cdot\|_\infty^{c_1,c_2}$ is an $\mathscr{R}_{a,p}$-generating norm. In particular, we obtain (5.5).   □

In the following let $c_1 = (2 + 2a^{-1})/p$, $c_2 = (2 + 2a)/p$; that is, c_1 and c_2 denote the unique weights associated to the weighted Tschebyschow norm that allows a representation of $\mathscr{R}_{a,p}$ according to (5.5). Due to Lemma 5.8 we may conclude that the distance $d_m(R(X,1))$ between a rectangle $R(X,1) = R_{c_1,c_2}(X,1)$ and a fixed point $A_m \in \mathscr{D}$ is concave in X on any convex region of $\mathscr{U}_i(A_m,1)$, $1 \leq i \leq 4$. For the sake of simplicity, we define an arrangement of the plane $\mathbb{R}^2$ such that each face of the arrangement is a subset of a region $\mathscr{U}_i(A_m,1)$, $1 \leq i \leq 4$. To this end we need the following notation.

Definition 5.11. Let $A_m \in \mathscr{D}$ be a fixed point. For each fundamental direction E of the Manhattan norm $\|\cdot\|_1$, straight lines with direction E through the points

$$A_m + \frac{1}{c_1 c_2} \begin{pmatrix} c_2 \\ c_1 \end{pmatrix}, \qquad\qquad A_m + \frac{1}{c_1 c_2} \begin{pmatrix} -c_2 \\ c_1 \end{pmatrix},$$

$$A_m + \frac{1}{c_1 c_2} \begin{pmatrix} c_2 \\ -c_1 \end{pmatrix}, \qquad\qquad A_m + \frac{1}{c_1 c_2} \begin{pmatrix} -c_2 \\ -c_1 \end{pmatrix}.$$

are called *auxiliary lines*.

Drawing all $4M$ auxiliary lines associated to the fixed points in $\mathscr{D}$ induces a partition of the plane $\mathbb{R}^2$ into convex regions. As in previous chapters, the arrangement which is induced by the auxiliary lines is called $\mathscr{A}(\mathscr{L})$. We assume that each face of the arrangement $\mathscr{A}(\mathscr{L})$ is closed. The following result shows that the objective function $f(R(X,1))$ of the rectangle location problem with feasible set $\mathscr{R}_{a,p}$ is concave in X on each face of the arrangement $\mathscr{A}(\mathscr{L})$.

Lemma 5.12. *Let $\mathscr{F}$ be a face of the arrangement $\mathscr{A}(\mathscr{L})$. Then the point–rectangle distance $d(R(X,1),A_m)$ between a rectangle $R(X,1) \in \mathscr{R}_{a,p}$ and a point $A_m \in \mathscr{D}$ is concave in X on $\mathscr{F}$. In particular, the objective function of the rectangle location problem with feasible set $\mathscr{R}_{a,p}$ is concave in X on $\mathscr{F}$.*

Proof. From the definition of auxiliary lines and the regions $\mathscr{U}_i(A_m,1)$ it follows that for each face $\mathscr{F}$ of the arrangement $\mathscr{A}(\mathscr{L})$ there exists a value $i_m \in \{1,\dots,4\}$ for all $m = 1,\dots,M$ such that

$$\mathscr{F} = \bigcap_{m=1}^{M} \mathscr{U}_{i_m}(A_m,1).$$

Thus, using Corollary 5.8 we may conclude that the distance $d(R(X,1),A_m)$ is concave on $\mathscr{F}$ for all $m = 1,\dots,M$. On each face $\mathscr{F}$ the objective function $f(R(X,1))$ is a sum of concave functions and therefore also concave in X. $\qquad\square$

Lemma 5.12 allows us to draw the following consequence.

Corollary 5.13. *There exists a minisum rectangle $R^* \in \mathscr{R}_{a,p}$ for the rectangle location problem with feasible set $\mathscr{R}_{a,p}$.*

Proof. On each face of the arrangement $\mathscr{A}(\mathscr{L})$ the objective function is concave and bounded below by zero. Therefore, there has to exist $X \in \mathbb{R}^2$ such that $R_{c_1,c_2}(X,1)$ is a minisum rectangle. $\qquad\square$

Note that there exists a vertex X of the arrangement $\mathscr{A}(\mathscr{L})$ such that $R_{c_1,c_2}(X,1)$ is a minisum rectangle. This observation implies an incidence property which is satisfied by at least one minisum rectangle.

Lemma 5.14. *There exists a minisum rectangle R^* for the rectangle location problem with feasible set $\mathscr{R}_{a,p}$ such that at least one vertex V of R^* is at an intersection point of two direction lines, i.e., V is the intersection point between two distinct fundamental directions of norm $\|\cdot\|_1$ through two (not necessary distinct) fixed points.*

Proof. Assume that $R^* = R_{c_1,c_2}(X^*, 1)$ is a minisum rectangle and X^* is a vertex of the arrangement $\mathcal{A}(\mathcal{L})$. Then, X^* lies at an intersection point of two auxiliary lines. Let us assume that these lines are associated to the fixed points $A_1 = (a_{11}, a_{12})$ and $A_2 = (a_{21}, a_{22})$. Without loss of generality, let us assume that X^* lies on a horizontal auxiliary line ℓ_1 associated to A_1 and a vertical auxiliary line ℓ_2 associated to A_2. We obtain $|a_{11} - x_1| = c_1^{-1}$ and $|a_{22} - x_2| = c_2^{-1}$ where $X^* = (x_1, x_2)$. In particular, we have $|y_1 - x_1| = c_1^{-1}$ for any $Y = (y_1, y_2) \in \ell_1$ and $|z_2 - x_2| = c_2^{-1}$ for any $Z = (z_1, z_2) \in \ell_2$, see Fig. 5.4. Let $Y' = (y_1', y_2')$ denote the intersection point between ℓ_1 and ℓ_2. Then we have $c_1|y_1' - x_1| = 1$ and $c_2|y_2' - x_2| = 1$; that is, Y' is a vertex of $R^* = R(X, 1)$. $\square$

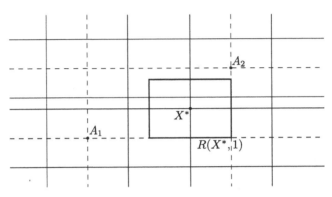

Fig. 5.4 Illustration of Lemma 5.14: auxiliary lines (*solid lines*) and direction lines (*dashed lines*) for the fixed points A_1 and A_2

Lemma 5.14 gives rise to a finite dominating set for the rectangle location problem with feasible set $\mathcal{R}_{a,p}$. At each intersection point V of two direction lines there exist four different rectangles $R \in \mathcal{R}_{a,p}$ having a vertex at V. The function Ξ which maps each instance of the rectangle location problem with feasible set $\mathcal{R}_{a,p}$ onto these rectangles is a finite dominating set according to Definition 1.56. There are $\mathcal{O}(M^2)$ intersection points between direction lines, therefore $\mathcal{O}(M^2)$ rectangles in $\Xi(\mathcal{I})$ for each instance $\mathcal{I}$. Hence, the brute-force algorithm which evaluates each rectangle $R \in \Xi(\mathcal{I})$ needs $\mathcal{O}(M^3)$ operations. Alternatively, it is also possible to evaluate each intersection point of two auxiliary lines. There are $\mathcal{O}(M^2)$ intersection points between auxiliary lines. Hence, also this approach leads to a minisum rectangle in $\mathcal{O}(M^3)$ operations.

Examples can be readily constructed that show the existence of minisum rectangles which do not have any vertex at an intersection point of two direction lines. Therefore, the finite dominating set Ξ does not contain all minisum rectangles in general. However, the set of optimal centers

$$\mathcal{X}^* := \{X \in \mathbb{R}^2 : f(R_{c_1,c_2}(X, 1)) \leq f(R_{c_1,c_2}(Y, 1)) \text{ for all } Y \in \mathbb{R}^2\}$$

has to be the union of vertices, edges, and faces of the arrangement $\mathscr{A}(\mathscr{L})$. Indeed, on each face $\mathscr{F}$ of the arrangement $\mathscr{A}(\mathscr{L})$ the objective function is concave. Thus, either all points of $X \in \mathscr{F}$, or all points X lying on an edge of $\mathscr{F}$, or a vertex X of $\mathscr{F}$ correspond to a local minisum rectangle $R(X,1)$. Therefore, we can find all minisum rectangles in the following way: Firstly, compute all minisum rectangles having their centers at vertices of $\mathscr{A}(\mathscr{L})$, and let f^* denote the objective value of a minisum rectangle. Secondly, for all edges E and faces $\mathscr{F}$ of $\mathscr{A}(\mathscr{L})$, choose any interior point $X \in E$ and $X \in \mathscr{F}$, respectively, and compute the objective value of the rectangle $R = R_{c_1,c_2}(X,1)$. If $f(R) = f^*$ then the edge E and the face $\mathscr{F}$, respectively, belong to $\mathscr{X}^*$. The number of edges and faces of the arrangement $\mathscr{A}(\mathscr{L})$ is bounded by $\mathscr{O}(M^2)$ (cf. [BKOS00]), therefore the worst case complexity of the brute-force algorithm mentioned above does not increase if we search all minisum rectangles.

5.4.2 Equal Aspect Ratio

The rectangle location problem with feasible set $\mathscr{R}_a$ is discussed in the following.

Lemma 5.15. *For any $a > 0$ the weighted Tschebyschow norm $\| \cdot \|_{\infty}^{1,a}$ is an $\mathscr{R}_a$-generating norm; that is,*

$$\mathscr{R}_a = \{R_{1,a}(X,r) : X \in \mathbb{R}^2, r > 0\}.$$

Proof. On the one hand, any rectangle $R_{1,a}(X,r)$ has aspect ratio a. Hence,

$$\{R_{1,a}(X,r) : X \in \mathbb{R}^2, r > 0\} \subseteq \mathscr{R}_a.$$

On the other hand, for any rectangle $R \in \mathscr{R}_a$ having its center at $X \in \mathbb{R}^2$ there exists $\lambda > 0$ such that

$$\lambda R_{1,a}(O,1) + \{X\} = R;$$

that is, $R_{1,a}(X,\lambda) = R$ and $\mathscr{R}_a \subseteq \{R_{c_1,c_2}(X,r) : X \in \mathbb{R}^2, r > 0\}$.   □

From Lemma 5.15 it follows that the rectangle location problem with feasible set $\mathscr{R}_{a,p}$ is equivalent to the minisum circle problem with unequal norms where the weighted Tschebyschow norm $\| \cdot \|_{\infty}^{1,a}$ is used in order to define circles and the Manhattan norm is used in order to define distances, see Chapter 4. Hence, there always exists a minisum rectangle $R^* \in \mathscr{R}_{a,p}$ (cf. Theorem 4.19) and the finite dominating set which has been obtained in Section 4.4.2 is also valid for the present rectangle location problem.

The following considerations lead to a geometric description of minisum rectangles. Firstly, we can state the following incidence property which is derived from Lemma 5.14.

Lemma 5.16. *There exists a minisum rectangle R^* for the rectangle location problem with feasible set $\mathcal{R}_a$ such that at least one vertex V of R^* is at an intersection point of two direction lines, i.e., V is the intersection point between two direction lines through two (not necessary distinct) fixed points.*

Proof. Due to Lemma 4.19 there exists a minisum rectangle $R^* \in \mathcal{R}_a$. Let p be its perimeter. Since R^* is an optimal solution to the rectangle location problem with feasible set $\mathcal{R}_a$ it has to be an optimal solution to the problem with feasible set $\mathcal{R}_{a,p}$. Using Lemma 5.14 we may conclude that there exists a rectangle $R' \in \mathcal{R}_{a,p}$ having a vertex at the intersection point of two direction lines through two fixed points such that $f(R') = f(R^*)$. Since $R' \in \mathcal{R}_a$ we obtain the assertion.   □

Remark 5.17. Recall that for the rectangle location problem with feasible set $\mathcal{R}_{a,p}$ there exists a minisum rectangle having its center at the intersection point of two auxiliary lines. This observation is not sufficient in order to show Lemma 5.16 due to the fact that it is not possible to define auxiliary lines for the rectangle location problem with feasible set $\mathcal{R}_a$. Therefore, Lemma 5.14 is crucial to the proof of Lemma 5.16.

The number of rectangles having a vertex at an intersection point between direction lines is not bounded. Nevertheless, we can show that only a finite number of such rectangles has to be considered in order to find a minisum rectangle. To this end consider the $\mathcal{R}_a$-generating norm $\| \cdot \|_\infty^{1,a}$ and let $V_1 = (v_1, v_2) \in \mathbb{R}^2$ be an arbitrary point. For any $z > 0$ let

$$X_z := (z + v_1, a^{-1}z + v_2)$$

and define the rectangle $R_z := R_{c_1, c_2}(X_z, z)$. Note that for all $z > 0$ we have $\|X_z - V_1\|_\infty^{1,a} = z$. Hence, for all $z > 0$ the bottom left vertex of R_z is at V_1. Furthermore, the bottom right, top right, and top left vertices are given by

$$V_2 = (v_1 + 2z, v_2), \qquad V_3 = (v_1 + 2z, v_2 + 2a^{-1}z), \qquad V_4 = (v_1, v_2 + 2a^{-1}z).$$

Remark 5.18. For any $a > 0$ the fundamental directions of the norm $\| \cdot \|_\infty^{1,a}$ are given as

$$E_1 = \begin{pmatrix} 1 \\ a^{-1} \end{pmatrix}, \qquad E_2 = \begin{pmatrix} 1 \\ -a^{-1} \end{pmatrix}, \qquad E_3 = \begin{pmatrix} -1 \\ -a^{-1} \end{pmatrix}, \qquad E_4 = \begin{pmatrix} -1 \\ a^{-1} \end{pmatrix}.$$

Hence, for any $z > 0$ we have $X_z = V_1 + zE_1$ and the center of $R_z = R_{1,a}(X_z, z)$ lies on the ray $[V_1, E_1\rangle$.

For each fixed point $A_m = (a_{m1}, a_{m2}) \in \mathcal{D}$ define

$$\pi_m^1 = \frac{a_{m1} - v_1}{2}, \qquad \pi_m^2 = \frac{a(a_{m2} - v_2)}{2}.$$

With these notations we can state the following result.

Lemma 5.19. *For any $V_1 \in \mathbb{R}^2$ and any fixed point $A_m \in \mathcal{D}$, the function $h(z) = d(R(X_z,z),A_m)$ is piecewise concave in z on $[0,\infty[$. It has a breakpoint in each point of the set $\{\pi_m^1, \pi_m^2\} \cap [0,\infty[$.*

Proof. Note that

$$\mathcal{S} = \{(X_z,z) : z \in [0,\infty[\} \subseteq \mathbb{R}^2 \times [0,\infty[$$

is a ray with start point in $(V,0)$. Combining Lemma 4.3 and Lemma 5.9 we obtain that the point–rectangle distance is piecewise concave on $\mathcal{S}$. Hence, $h(z)$ has to be piecewise concave in z on $[0,\infty[$.

The point–rectangle distance $d_m(R(X,r))$ has a breakpoint in $(X,r) \in \mathcal{S}$ if and only if (X,r) belongs to two 3-faces of the arrangement $\mathcal{A}^+(\mathcal{H}_m)$. This implies that $h(z)$ has a breakpoint in z' if and only if $A_m \in U_{i_1}(X_{z'},z')$ and $A_m \in U_{i_2}(X_{z'},z')$ for some $i_1 \neq i_2 \in \{1,2,3,4\}$. Due to the definition of the regions $\mathcal{U}_{i_1}(X_{z'},z')$ and $\mathcal{U}_{i_2}(X_{z'},z')$ it follows that either the vertical line $\langle V_2,V_3 \rangle$ through the vertices V_2 and V_3 of $R_z = R(X_{z'},z')$ or the horizontal line $\langle V_3,V_4 \rangle$ through the vertices V_3 and V_4 intersect A_m. In the former case we have $z' = \pi_m^2$ and in the latter case we have $z' = \pi_m^1$. $\qquad\square$

Using Lemma 5.19 we obtain the following consequence.

Corollary 5.20. *For any $V_1 \in \mathbb{R}^2$, the function*

$$h(z) = f(R(X_z,z)) = \sum_{m=1}^{M} \omega_m d(R(X_z,z),A_m)$$

is piecewise concave in z on $[0,\infty[$. It has a breakpoint in each point of the set $\{\pi_m^1, \pi_m^2 : 1 \leq m \leq M\} \cap [0,\infty[$.

Since we may rotate the fixed points by multiples of $90°$, a similar result to Corollary 5.20 applies if we keep another vertex than the bottom left vertex of the rectangle R_z fixed. Furthermore, $z = \pi_m^i$, $i = 1,2$, implies that an edge of $R(X_z,z)$ is contained in a direction line through the fixed point A_m. Hence, we have the following incidence property for the objective function of the rectangle location problem with feasible set $\mathcal{R}_a$.

Lemma 5.21. *There exists a minisum rectangle R^* for the rectangle location problem with feasible set $\mathcal{R}_a$ such that a vertex of R^* is at an intersection point of two direction lines $\ell_1 \neq \ell_2$ and an edge of R^* is contained in a third direction line ℓ_3 distinct from ℓ_1 and ℓ_2.*

Remark 5.22. Note that Lemma 5.21 is an enhancement of a similar result for the minisum circle problem with unequal norms stated in Lemma 4.22.

For any intersection point of two direction lines there exist at most $8(M-1)$ rectangles fulfilling the incidence property of Lemma 5.21. Hence, the size of set

$\Xi(I)$ which contains all rectangles with this incidence property is $\mathcal{O}(M^3)$ for any instance $I \in \mathscr{I}$ of the rectangle location problem with feasible set $\mathscr{R}_a$, and the function $\Xi : I \mapsto \Xi(I)$ is a finite dominating set for this problem. Using a brute-force algorithm a minisum rectangle may be computed in $\mathcal{O}(M^4)$ operations.

As for the problem with feasible set $\mathscr{R}_{a,p}$ also for the problem with feasible set $\mathscr{R}_a$ it can be shown that Lemma 5.21 does not state a necessary condition. In order to find all minisum rectangles having a vertex at an intersection point of two direction lines it has to be checked whether or not the objective value $f(R_z)$ is constant in z on a line segment defined by the values $\pi_m^1, \pi_m^2, 1 \leq m \leq M$. Algorithm 5.1 evaluates all rectangles contained in the finite dominating set $\Xi(I)$ and also reports minisum rectangles which have a vertex at an intersection point of two direction lines but do not satisfy the third incidence property of Lemma 5.21. Using a comparison-based sorting procedure the algorithm runs in $\mathcal{O}(M^4 \log M)$ operations.

Algorithm 5.1: Computing minisum rectangles for feasible set $\mathscr{R}_a$

Input: Set of fixed points $\mathscr{D}$, associated weights ω_m, aspect ratio $a > 0$, weights c_1, c_2 of R-generating norm for $\mathscr{R}_a$

Output: Set of all minisum rectangles Opt with vertex at an intersection point of direction lines

```
   // Initialization
1  Opt := ∅, X* := (0,0), r* := 1, R* := R_{c₁,c₂}(X*,r*)
   // Search for minisum rectangles
2  for any intersection point V of two direction lines do
3  │   for φ = 0°,90°,180°,270° do
4  │   │   Rotate 𝒟 by φ degree in clockwise direction around the origin
5  │   │   for m = 1,...M do
6  │   │   │   Compute πₘ¹ and πₘ² associated to V
7  │   │   │   if πₘⁱ ≤ 0 then
8  │   │   │   └   Delete πₘⁱ
9  │   │   Sort the remaining values πₘⁱ in an increasing order, adapt M
10 │   │   Rename the values as μⱼ, 1 ≤ j ≤ 2M, such that μⱼ ≤ μⱼ₊₁
11 │   │   for 1 ≤ j ≤ 2M do
   │   │   │   // Rectangles contained in the FDS
12 │   │   │   Rⱼ := R_{c₁,c₂}(X_{μⱼ},r_{μⱼ})
13 │   │   │   fⱼ := f(Rⱼ)
14 │   │   │   if fⱼ = f(R*) then Opt := Opt ∪ {Rⱼ}
15 │   │   │   if fⱼ < f(R*) then
16 │   │   │   │   Opt := {Rⱼ}
17 │   │   │   └   X* := X_{μⱼ}, r* := r_{μⱼ}
   │   │   │   // Minisum rectangles not contained in the FDS
18 │   │   │   if Rⱼ₋₁ ∈ Opt then
19 │   │   │   │   z := ½(μⱼ₋₁ + μⱼ), R_z := R_{c₁,c₂}(X_z,z)
20 │   │   │   └   if f(R_z) = f(R*) then Opt := Opt ∪ {R_z : z ∈ [μⱼ₋₁,μⱼ]}
```

Remark 5.23. Small modifications of Algorithm 5.1 lead to an $\mathcal{O}(M^3 \log M)$ algorithm: As mentioned above, $f(R_z)$ is concave in z on intervals which are defined by the values π_m^1, π_m^2, $1 \leq m \leq M$. Decomposing each interval into $\mathcal{O}(1)$ smaller intervals results in an affine linear function on each interval. Now, the evaluation of $f(R_j)$ in the fourth *for loop* of Algorithm 5.1 can be performed in $\mathcal{O}(M)$ operations and the runtime reduces to $\mathcal{O}(M^3 \log M)$.

Remark 5.24. Note that the finite dominating set which is derived in Chapter 4 consists of $\mathcal{O}(M^3)$ candidate circles (i.e., rectangles in our case). Thus, the finite dominating set for the rectangle location problem with feasible set $\mathcal{R}_a$ has the same size but gives rise to an efficient enumeration scheme.

5.4.3 Equal Perimeter

Given a fixed perimeter $p > 0$ we study the rectangle location problem with feasible set $\mathcal{R}_p$. For analyzing this problem let $V = (v_1, v_2) \in \mathbb{R}^2$ be any point and define

$$X_{V,z} := \begin{pmatrix} 2^{-1}z + v_1 \\ 4^{-1}(p - 2z) + v_2 \end{pmatrix}, \qquad c_1(z) := \frac{2}{z}, \qquad c_2(z) := \frac{4}{p - 2z}.$$

Then it can be readily verified that for any $0 < z < 2^{-1}p$

$$R_{V,z} = \{Y \in \mathbb{R}^2 : \|Y - X_{V,z}\|_\infty^{c_1(z),c_2(z)} = 1\}$$

is a rectangle with perimeter p having its bottom left vertex at point V and its center at $X_{V,z}$. Furthermore, the length of both horizontal edges of the rectangle $R_{V,z}$ is z. As a consequence we obtain the following result.

Lemma 5.25. *For any $p > 0$ we have*

$$\mathcal{R}_p = \bigcup_{V \in \mathbb{R}^2} \{R_{V,z} : 0 < z < 2^{-1}p\}.$$

Remark 5.26. A rectangle $R_{V,z}$ can be represented by using the weighted Tschebyschow norm $\|Y - X_{V,z}\|_\infty^{c_1(z),c_2(z)}$. It is important to keep in mind that we have $R_{V,z} = R_{c_1(z),c_2(z)}(X_z, 1)$. In particular, each rectangle $R_{V,z}$ has radius $r = 1$. In contrast to previous rectangle location problems the radius is not used in order to define the side length of the rectangle $R_{V,z}$.

Remark 5.27. Note that the representation of $\mathcal{R}_p$ stated in Lemma 5.25 is based on different weighted Tschebyschow norms. In contrast to the sets $\mathcal{R}_{a,p}$ and $\mathcal{R}_a$, it is clear that an $\mathcal{R}_p$-generating norm does not exist.

The following example shows that a minisum rectangle for the rectangle location problem with feasible set $\mathcal{R}_p$ does not need to exist.

Example 5.28. Suppose $M = 2$, $A_1 = (0,0)$, $A_2 = (0,10)$, $\omega_1 = \omega_2$, and $p = 20$. Let $V = A_1$ and consider the rectangle $R_{V,z}$ having its bottom left vertex in V. We have

$$f(R_{V_z}) = \omega_1 v_1 + \omega_2(10 - v_1 - z) = \omega_2(10 - z).$$

This function attains on $[0, 2^{-1}p]$ a minimum in $z = 2^{-1}p = 10$. This value does not correspond to a rectangle but to a vertical line segment with length $2^{-1}p$. Since $f(R_{V,z}) = 0$ for $V = A_1$, $z = 10$ and $f(R_{V,z}) > 0$ for all $V \neq A_1$ it can be concluded that $\mathcal{R}_p$ does not contain a minisum rectangle in this case.

In order to ensure the existence of a minimum for the rectangle location problem with feasible set $\mathcal{R}_p$ the following definition is needed.

Definition 5.29. The set containing $\mathcal{R}_p$ and all vertical and horizontal line segments with length $p/2$ is called $\overline{\mathcal{R}}_p$. The elements of $\overline{\mathcal{R}}_p$ are called *generalized rectangles*.

We show that $\overline{\mathcal{R}}_p$ always contains an optimal solution to the rectangle location problem with feasible set $\overline{\mathcal{R}}_p$.

Lemma 5.30. *For any $p > 0$ there exists an optimal solution $R \in \overline{\mathcal{R}}_p$ to the rectangle location problem with feasible set $\overline{\mathcal{R}}_p$.*

Proof. Let $R_{\mathcal{D}}$ denote the smallest rectangle enclosing $\mathcal{D}$. Let

$$\mathcal{B} := \{V \in \mathbb{R}^2 : \min_{Y \in R_{\mathcal{D}}} \|V - Y\|_\infty \leq p/2\}$$

and define for each $V \in \mathcal{B}$ the set $\mathcal{R}_V \subseteq \overline{\mathcal{R}}_p$ which consists of all generalized rectangles having perimeter p and bottom left vertex at V. Denoting the length along the x-axis of a generalized rectangle by z we may identify the set $\mathcal{R}_V$ by $\{V\} \times [0, 2^{-1}p]$. Therefore, we may conclude that

$$\overline{\mathcal{R}}'_p := \bigcup_{V \in \mathcal{B}} \mathcal{R}_V$$

is a compact set in $\mathbb{R}^3$, and the objective function of the rectangle location problem with feasible set $\overline{\mathcal{R}}'_p$ attains a minimum on $\overline{\mathcal{R}}'_p$. It remains to show that $\overline{\mathcal{R}}_p \setminus \overline{\mathcal{R}}'_p$ does not contain a generalized minisum rectangle. To this end, note that $\min_{Y \in R_{\mathcal{D}}} d(R,Y) > 0$ for all $R \in \overline{\mathcal{R}}_p \setminus \overline{\mathcal{R}}'_p$. Hence, $R \in \overline{\mathcal{R}}_p \setminus \overline{\mathcal{R}}'_p$ does not intersect $R_{\mathcal{D}}$ and we can improve the objective value of R by moving R toward $R_{\mathcal{D}}$.   □

Similar to rectangle location problems with feasible sets $\mathcal{R}_{a,p}$ or $\mathcal{R}_a$ we have the following incidence property.

Lemma 5.31. *There exists a (degenerated) minisum rectangle R^* for the rectangle location problem with feasible set $\overline{\mathcal{R}}_p$ such that at least one vertex V of R^* is at an intersection point of two distinct direction lines.*

Proof. Let $R^* \in \overline{\mathscr{R}_p}$ be a minisum rectangle. We distinguish the case $R^* \in \mathscr{R}_p$ and $R^* \in \overline{\mathscr{R}_p} \setminus \mathscr{R}_p$. In the former case the representation

$$\mathscr{R}_p = \bigcup_{a>0} \mathscr{R}_{a,p}$$

implies the assertion. In the latter case R^* is either a vertical or a horizontal line segment with length $p/2$. We denote as ℓ the straight line containing R^*. Furthermore, we denote as L the stripe consisting of all straight lines orthogonal to ℓ and intersecting R^*. Let $H_\ell^+, H_\ell^- \subseteq \mathbb{R}^2$ denote both open half-spaces defined by ℓ. Also the stripe L defines two open half-spaces; these half-spaces are called H_L^+ and H_L^-. It can be readily shown that R^* satisfies the following median properties:

$$\left| \sum_{m:A_m \in H_\Psi^+} \omega_m - \sum_{m:A_m \in H_\Psi^-} \omega_m \right| \leq \sum_{m:A_m \in \Psi} \omega_m, \quad \Psi \in \{\ell, L\}.$$

Hence, an optimal degenerated minisum rectangle R' exists such that both the corresponding line ℓ and the boundary of the corresponding stripe L contain at least one fixed point. It follows that a vertex of R' lies at an intersection point of two direction lines of two distinct fixed points. □

By using Lemma 5.30 and Lemma 5.31 we may conclude that there exists an intersection point of two direction lines V and $\phi \in \{0°, 90°, 180°, 270°\}$ such that

- Either $R_{V,z}$ is a minisum rectangle for some $z \in]0, 2^{-1}p[$ where the fixed points are rotated around the origin by ϕ, or
- One of the four vertical and horizontal line segments with start point in V and length $2^{-1}p$ is a degenerated minisum rectangle

The following result is needed in order to prove a stronger incidence property which applies to some minisum rectangles $R^* \in \mathscr{R}_p$. Note that the following result does not follow directly from properties of the point–rectangle distance since there does not exist an $\mathscr{R}_p$-generating norm.

Lemma 5.32. *Let* $V = (v_1, v_2) \in \mathbb{R}^2$. *Then the function* $h(z) = f(R_{V,z})$ *is piecewise concave in* z *on* $]0, p/2[$. $h(z)$ *has a breakpoint in each point of the set*

$$\{\pi_m^1, \pi_m^2 : 1 \leq m \leq M\} \cap]0, p/2[$$

where $\pi_m^1 = a_m 1 - v_1$ *and* $\pi_m^2 = a_m 2 - v_2 - p/2$, $1 \leq m \leq M$.

Proof. We show that the distance $d_m(R_{V,z}, A_m)$ between a fixed point $A_m \in \mathscr{D}$ and a rectangle $R_{V,z}$ is piecewise concave in z on $]0, p/2[$ and has breakpoints in $\{\pi_m^1, \pi_m^2\} \cap]0, p/2[$. This shows the result, since $h(z) = f(R_{V,z})$ is the sum of weighted distances $d_m(R_{V,z})$ where all weights are nonnegative.

First, it should be noted that for each value of $z \in]0, p/2[$ norm $\| \cdot \|_{\infty}^{c_1(z), c_2(z)}$ is used in order to define region $\mathscr{U}_i(X_z, 1)$. In order to emphasize the relation between norm $\| \cdot \|_{\infty}^{c_1(z), c_2(z)}$ and region $\mathscr{U}_i(X_z, 1)$ we use in the following the denotation $\mathscr{U}_i^{c_1(z), c_2(z)}(X_z, 1)$.

Without loss of generality assume $\pi_m^1 \leq \pi_m^2$, otherwise we may change the numeration of the values. Since $\pi_m^1, \pi_m^2 \in \mathbb{R}$ three cases are possible:

(i) $\pi_m^1, \pi_m^2 \notin]0, p/2[\cap [\pi_m^1, \pi_m^2]$.
(ii) $\pi_m^i \in]0, p/2[\cap [\pi_m^1, \pi_m^2]$, $\pi_m^j \notin]0, p/2[\cap [\pi_m^1, \pi_m^2]$, $i \neq j \in \{1, 2\}$.
(iii) $\pi_m^1, \pi_m^2 \in]0, p/2[\cap [\pi_m^1, \pi_m^2]$.

Due to the definition of π_m^1 and π_m^2 case (i) implies that there exists $j \in \{1, 2, 3\}$ such that

$$A_m \in \mathscr{U}_j^{c_1(z), c_2(z)}(X_z, 1) \text{ for all } z \in]0, p/2[.$$

Case (ii) implies that there exist $j_1 \neq j_2 \in \{1, 2, 3\}$ such that

$$A_m \in \mathscr{U}_{j_1}^{c_1(z), c_2(z)}(X_z, 1) \text{ for all } z \in]0, \pi_m^i]$$

and

$$A_m \in \mathscr{U}_{j_2}^{c_1(z), c_2(z)}(X_z, 1) \text{ for all } z \in [\pi_m^i, p/2[$$

where $i \in \{1, 2\}$ such that $\pi_m^i \in]0, p/2[$. Case (iii) implies that there exist $j_1 \neq j_2 \in \{1, 2, 3\}$ such that

$$A_m \in \mathscr{U}_{j_1}^{c_1(z), c_2(z)}(X_z, 1) \text{ for all } z \in]0, \pi_m^1],$$

$$A_m \in \mathscr{U}_4^{c_1(z), c_2(z)}(X_z, 1) \text{ for all } z \in [\pi_m^1, \pi_m^2],$$

and

$$A_m \in \mathscr{U}_{j_2}^{c_1(z), c_2(z)}(X_z, 1) \text{ for all } z \in]\pi_m^2, p/2[.$$

Since $c_1(z) = 2/z$ and $c_2(z) = 4/(p - 2z)$, it follows from the formula for the point–rectangle distance stated in Lemma 5.6 that $d_m(R_{V,z})$ is piecewise concave with breakpoints in $\{\pi_m^1, \pi_m^2\} \cap]0, p/2[$. $\qquad\square$

Lemma 5.32 gives rise to the following incidence property.

Lemma 5.33. *There exists a generalized minisum rectangle $R^* \in \overline{\mathscr{R}_p}$ for the rectangle location problem with feasible set $\mathscr{R}_p$ such that a vertex of R^* is at an intersection point of two distinct direction lines $\ell_1 \neq \ell_2$ and an edge of R^* is contained in a third direction line ℓ_3 which is not necessary distinct from ℓ_1 or ℓ_2. In particular, if ℓ_3 is distinct from ℓ_1 and ℓ_2, then R^* is nondegenerated.*

Proof. From Lemma 5.31 we obtain that there exists a (degenerated) minisum rectangle having a vertex at an intersection point of two direction lines. Without loss of generality we assume that a (degenerated) minisum rectangle R^* exists having its bottom left vertex at an intersection point of two direction lines, say V. Since $h(z) = f(R_{V,z})$ is piecewise concave in z on $]0, p/2[$ it follows that there also exists a (degenerated) minisum rectangle R' such that $R' = R_{V,z}$ for some

$$z \in \{0, p/2\} \cup \left(\{ \pi_m^1, \pi_m^2 : 1 \leq m \leq M \} \cap]0, p/2[\right).$$

$z \in \{0, p/2\}$ implies that R' is a degenerated minisum rectangle. Thus, the start point V of the line segment R' lies at an intersection point of two direction lines ℓ_1, ℓ_2 and R' is contained either in ℓ_1 or ℓ_2. If $z = \pi_m^i$ for some $m \in \{1, \ldots, M\}$ and $i \in \{1, 2\}$ it follows that R' is not degenerated. In this case $i = 1$ implies that the horizontal edge of R' which is not adjacent to V is contained in the horizontal direction line through A_m. Similarly, $i = 2$ implies that the vertical edge of R' which is not adjacent to V is contained in the vertical direction line through A_m.   □

Lemma 5.33 gives rise to a solution method for the rectangle location problem with feasible set $\overline{\mathcal{R}_p}$. There are $\mathcal{O}(M^3)$ rectangles $R \in \mathcal{R}_p$ satisfying the conditions of Lemma 5.33. Furthermore, it can be concluded that there exist at most $2M(M-1)$ line segments satisfying the conditions of the lemma. Thus, an optimal generalized minisum rectangle may be found in $\mathcal{O}(M^4)$ operations by evaluating $\mathcal{O}(M^3)$ generalized rectangles. As a solution procedure a two-step approach is suitable. In a first step, Algorithm 5.2 may be used in order to compute optimal nondegenerated rectangles $R \in \mathcal{R}_p$ having a vertex at an intersection point of two direction lines. In a second step, degenerated rectangles $R \in \overline{\mathcal{R}_p} \setminus \mathcal{R}_p$ have to be evaluated. The worst-case complexity of Algorithm 5.2 is $\mathcal{O}(M^4)$. Evaluating all candidate line segments can be performed in $\mathcal{O}(M^3)$ operations by a brute-force algorithm. Hence, the following result may be obtained.

Lemma 5.34. *All generalized minisum rectangles with perimeter $p > 0$ having a vertex at an intersection point of two direction lines may be computed in $\mathcal{O}(M^4)$ operations.*

Remark 5.35. As we have already mentioned in the proof of Lemma 5.31 a degenerated rectangle has to satisfy the median properties

$$\left| \sum_{m: A_m \in H_\Psi^+} \omega_m - \sum_{m: A_m \in H_\Psi^-} \omega_m \right| \leq \sum_{m: A_m \in \Psi} \omega_m, \quad \Psi \in \{\ell, L\}.$$

Hence, an optimal degenerated rectangle can be computed in $\mathcal{O}(M \log M)$ operations.

For equal weighted fixed points the following result states a mild condition ensuring that at least one nondegenerated minisum rectangle exists.

Theorem 5.36. *Suppose that all fixed points in $\mathcal{D}$ have equal weight $\omega > 0$ and no pair of two fixed points lies on a common vertical or horizontal line. Then there exists a nondegenerated minisum rectangle $R \in \mathcal{R}_p$ for the rectangle location problem with feasible set $\mathcal{R}_p$.*

Algorithm 5.2: Computing optimal rectangles $R \in \mathscr{R}_p$

Input: Set of fixed point $\mathscr{D}$, associated weights ω_m, perimeter $p > 0$
Output: Set of optimal rectangles Opt $\subseteq \mathscr{R}_p$ with vertex at an intersection point of direction
   lines
```
// Initialization
```
1 Opt $:= \emptyset$, $f^* := \infty$
```
// Search for minisum rectangles
```
2 **for** *any intersection point V of two direction lines* **do**
3   **for** $\phi = 0°, 90°, 180°, 270°$ **do**
4   Rotate $\mathscr{D}$ by ϕ degree in clockwise direction around the origin
5   **for** $m = 1, \ldots M$ **do**
6   Compute π_m^1 and π_m^2 associated to V
7   **if** $\pi_m^i \notin \,]0, 2^{-1}p[$ **then**
8   Delete π_m^i

9   Let M' be the number of remaining values π_m^i
10   **for** $1 \leq j \leq M'$ **do**
11           $f_j := R_{V,\mu_j}$
12   **if** $f_j = f^*$ **then**
13   Opt $:=$ Opt $\cup \{R_{V,\mu_j}\}$
14   **if** $f_j < f^*$ **then**
15   Opt $:= \{R_{V,\mu_j}\}$
16              $f^* := f_j$

Proof. Assume that $S \subseteq \mathbb{R}^2$ is a degenerated minisum rectangle, i.e., S is either a horizontal or a vertical line segment with length $2^{-1}p$. We show that a nondegenerated rectangle $R \in \mathscr{R}_p$ exists such that $f(R) \leq f(S)$. To this end, assume without loss of generality:

- $\omega_m = 1$ for all $1 \leq m \leq M$.
- S is a horizontal line segment, and $S = [X, Y]$ for some $X, Y \in \mathbb{R}^2$.
- X is an intersection point of two distinct direction lines.
- Y is not a fixed point, i.e., $Y \notin \mathscr{D}$.
- $A_1 \in \mathscr{D}$ belongs to the straight line containing S.

Let $R_1(z) \in \mathscr{R}_p$ denote the rectangle with bottom left vertex in X and bottom right vertex in $X + (z, 0)$. Let $R_2(z)$ denote the rectangle with top right vertex in Y and top left vertex in $Y - (z, 0)$. Note that $R_1(z)$ and $R_2(z)$ are well defined for $z \in \,]0, 2^{-1}p[$. Denote the bottom left vertex of $R_i(z)$ as V_1^i, the bottom right vertex by V_2^i, the top right vertex by V_3^i, and the top left vertex by V_4^i, $i = 1, 2$. Then we obtain

$$V_1^1 = \begin{pmatrix} x_1 \\ x_2 \end{pmatrix}, \qquad\qquad V_2^1 = \begin{pmatrix} z + x_1 \\ x_2 \end{pmatrix},$$

$$V_3^1 = \begin{pmatrix} z + x_1 \\ -z + p/2 + x_2 \end{pmatrix}, \qquad V_4^1 = \begin{pmatrix} x_1 \\ -z + p/2 + x_2 \end{pmatrix},$$

$$V_1^2 = \begin{pmatrix} -z + y_1 \\ z + y_2 - p/2 \end{pmatrix}, \qquad V_2^2 = \begin{pmatrix} y_1 \\ z + y_2 - p/2 \end{pmatrix},$$

$$V_3^2 = \begin{pmatrix} y_1 \\ y_2 \end{pmatrix}, \qquad V_4^2 = \begin{pmatrix} -z + y_1 \\ y_2 \end{pmatrix},$$

where $X = (x_1, x_2)$ and $Y = (y_1, y_2)$. Observe that V_3^1 lies for all $z \in]0, 2^{-1}p[$ on a straight line with slope -1. Analogously, V_1^2 lies for all $z \in]0, 2^{-1}p[$ on a straight line with slope 1, see also Fig. 5.5.

We subdivide $\mathscr{D}$ into three sets:

$$\mathscr{D}_{\mathrm{I}} = \{A_m = (a_{m1}, a_{m2}) \in \mathscr{D} : a_{m1} \leq x_1, a_{m2} > x_2\},$$
$$\mathscr{D}_{\mathrm{II}} = \{A_m = (a_{m1}, a_{m2}) \in \mathscr{D} : a_{m1} \geq y_1, a_{m2} < y_2\},$$
$$\mathscr{D}_{\mathrm{III}} = \mathscr{D} \setminus (\mathscr{D}_{\mathrm{I}} \cup \mathscr{D}_{\mathrm{II}} \cup \{A_1\}).$$

For any fixed point $A_m \in \mathscr{D}_{\mathrm{III}}$ there exists $\varepsilon_m^1 > 0$ such that

$$d(S, A_m) \geq d(R_1(z), A_m), \quad \text{for all } z \in]p/2 - \varepsilon_m^1, p/2[.$$

Indeed, if A_m lies above or below S then it is clear that the distance between A_m and $R_1(z)$ is either constant or decreases. If we have $a_{m1} \leq x_1$ and $a_{m2} < x_2$ then the distance between $A_m = (a_{m1}, a_{m2})$ and $R_1(z)$ is given by $\|X - A_m\|_1$; that is, the distance is constant for all $z \in]p/2 - \varepsilon_m^1, p/2[$. If $a_{m1} \geq y_1$ and $a_{m2} > y_2$ then the distance between A_m and $R_1(z)$ is given by $\|A_m - V_3^1\|$. Since V_3^1 lies on a line with slope -1 which intersects the point Y, we may conclude that the distance is constant for ε_m^1 sufficiently small. Analogously, we may show the same result for the distances between the fixed points $A_m \in \mathscr{D}_{\mathrm{III}}$ and the rectangle $R_2(z)$, i.e., for any fixed point $A_m \in \mathscr{D}_{\mathrm{III}}$ there exists $\varepsilon_m^2 > 0$ such that

$$d(S, A_m) \geq d(R_2(z), A_m), \quad \text{for all } z \in]p/2 - \varepsilon_m^2, p/2[.$$

For a fixed point $A_m \in \mathscr{D}_I$ there also exist $\varepsilon_m^1 > 0$ and $\varepsilon_m^2 > 0$ such that

$$d(S, A_m) - d(R_1(z), A_m) = \|A_m - X\| - \|A_m - V_4^1\|_1 = \frac{p}{2} - z,$$
$$d(S, A_m) - d(R_2(z), A_m) = \|A_m - X\| - \|A_m - V_4^2\|_1 = -\frac{p}{2} + z$$

for all $z \in]p/2 - \min\{\varepsilon_m^1, \varepsilon_m^2\}, p/2[$. Analogously, for a fixed point $A_m \in \mathscr{D}_{\mathrm{II}}$ there exist $\varepsilon_m^1 > 0$ and $\varepsilon_m^2 > 0$ such that

$$d(S, A_m) - d(R_1(z), A_m) = \|A_m - Y\| - \|A_m - V_2^1\|_1 = \frac{p}{2} + z,$$
$$d(S, A_m) - d(R_2(z), A_m) = \|A_m - Y\| - \|A_m - V_2^2\|_1 = \frac{p}{2} - z$$

for any $z \in]p/2 - \min\{\varepsilon_m^1, \varepsilon_m^2\}, p/2[$. If $A_1 \in]X, Y[$ then there exists $\varepsilon_1^1 > 0$ and $\varepsilon_1^2 > 0$ such that

$$d(S, A_1) = d(R_1(z), A_1) = d(R_2(z), A_1)$$

for all $z \in]p/2 - \min\{\varepsilon_1^1, \varepsilon_1^2\}, p/2[$. If $a_{11} \le x_1$ then we have

$$d(S, A_1) - d(R_1(z), A_1) = 0,$$
$$d(S, A_2) - d(R_2(z), A_1) \|A_1 - X\|_1 \|A_1 - V_4^2\|_1 = \frac{p}{2} - z$$

for all $z \in]0, p/2[$. Analogously, in case that $a_{11} \ge y_1$ we obtain

$$d(S, A_1) - d(R_1(z), A_1) = \|A_1 - Y\|_1 \|A_1 - V_2^2\|_1 = \frac{p}{2} - z,$$
$$d(S, A_2) - d(R_2(z), A_1) = 0$$

for all $z \in]0, p/2[$.

Let $\varepsilon := \min\{\varepsilon_m^i : m = 1, \ldots, M, \ i = 1, 2\}$. Then we obtain the following inequalities valid for any $z \in [p/2 - \varepsilon, p/2[$:

$$f(S) - f(R_1(z)) \ge (|\mathscr{D}_I| - |\mathscr{D}_{II}| - \sigma)\left(\frac{p}{2} - z\right)$$
$$f(S) - f(R_2(z)) \ge (|\mathscr{D}_{II}| - |\mathscr{D}_I| - \tau)\left(\frac{p}{2} - z\right)$$

where

$$\sigma = \begin{cases} 1; & a_{11} \ge y_1 \\ 0; & a_{11} < y_1 \end{cases}, \qquad \tau = \begin{cases} 1; & a_{11} \le x_1 \\ 0; & a_{11} > x_1 \end{cases}.$$

We have $\max\{(|\mathscr{D}_I| - |\mathscr{D}_{II}| - \sigma), (|\mathscr{D}_{II}| - |\mathscr{D}_I| - \tau)\} \ge 0$, hence there exists a non-degenerated minisum rectangle. $\qquad \square$

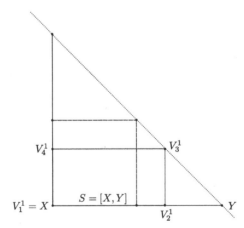

Fig. 5.5 Illustration of Theorem 5.36. The *vertical line segment* corresponds to $R_1(0)$, the *dashed rectangle* corresponds to $R_1(p/2)$, the *solid rectangle* corresponds to $R_1(3p/2)$, and the *horizontal line segment* $S = [X, Y]$ corresponds to $R_1(p/2)$

Remark 5.37. In the proof of Theorem 5.36 the observation is used that the top right vertex of the rectangle $R_1(z)$ lies on a straight line with slope -1. This observation can be generalized in the following way:

For two opposite vertices V and V' of an arbitrary rectangle $R \in \mathcal{R}_p$ we have

$$V' \in C(V, p/2), \quad V \in C(V', p/2) \tag{5.6}$$

where C is a circle defined with respect to the Manhattan norm.

The generalization of Theorem 5.36 to fixed points with distinct weights is not true. This result is shown by the following example.

Example 5.38. Let $p = 4$ and consider the four fixed points

$$A_1 = \begin{pmatrix} 1 \\ 0 \end{pmatrix}, \qquad A_2 = \begin{pmatrix} 0 \\ 1 \end{pmatrix}, \qquad A_3 = \begin{pmatrix} 3.5 \\ -0.5 \end{pmatrix}, \qquad A_4 = \begin{pmatrix} 3 \\ 0.5 \end{pmatrix}$$

with weights $\omega_1 = 4$, $\omega_2 = 1$, $\omega_3 = 2$, and $\omega_4 = 1$. Let R^i, $i = 1, 2, 3$, denote the three rectangles depicted in Fig. 5.6. The objective values of these rectangles are

$$f(R^1) = 6, \quad f(R^2) = 5, \quad f(R^3) = 5.$$

The horizontal line segment $S = [1, 3] \times \{0\}$ has objective value $\underline{f(S) = 4.5}$. Using Lemma 5.33 it can be verified that $f(S) \leq f(R)$ for all $R \in \overline{\mathcal{R}_p}$; that is, S is a unique optimal solution.

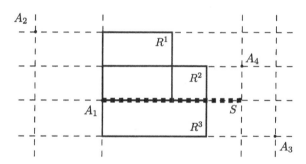

Fig. 5.6 Illustration of Example 5.38

5.5 Unrestricted Problem

In this section the case is considered where the feasible set is $\mathcal{R}$; that is, all axis-parallel rectangles are feasible solutions.

The following considerations show that the set of minisum rectangles for the rectangle location problem with feasible set $\mathcal{R}$ cannot contain only degenerated rectangles. Obviously, degenerated rectangles include points as well as horizontal and vertical linear curves (line segments, half-lines, straight lines). Besides that, also the following objects are degenerated rectangles:

- Bent lines composed of a vertical and a horizontal half-line
- Bent lines composed of two vertical half-lines and a horizontal line segment and vice versa
- Two parallel vertical or horizontal straight lines

The following result shows that a nondegenerated minisum rectangle always exists.

Lemma 5.39. *There always exists a minisum rectangle* $R^* \in \mathcal{R}$. *In particular, a point cannot be a degenerated minisum rectangle and a degenerated rectangle cannot be a unique optimal solution.*

Proof. Let $\tilde{R}$ be a degenerated rectangle. For each fixed point $A_m \in \mathcal{D}$ let $Y_m \in \tilde{R}$ such that $d_m(\tilde{R}) = \|Y - A_m\|_1$. Note that a nondegenerated rectangle $R \in \mathcal{R}$ exists such that $Y_m \in R$ for all $1 \leq m \leq M$. Hence, we have $f(R) \leq f(\tilde{R})$ and a degenerated minisum rectangle cannot be a unique optimal solution. This implies that a non-degenerated minisum rectangle $R \in \mathcal{R}$ exists. Furthermore, using Lemma 4.7 and the fact that $\mathcal{R} = \bigcup_{a>0} \mathcal{R}_a$ we may conclude that a point cannot be a degenerated minisum rectangle. $\qquad\square$

The following result states an incidence property for nondegenerated minisum rectangles.

Theorem 5.40. *There exists an optimal rectangle* $R \in \mathcal{R}$ *for the rectangle location problem with feasible set* $\mathcal{R}$ *such that each vertex of R lies at an intersection point of two direction lines.*

Proof. Due to Lemma 5.39 we may conclude that there exists an optimal rectangle, say $R^* = \partial(([x_1, x_2] \times [y_1, y_2])$. Let

$$V_0 = (x_1, y_2), \qquad V_1 = (x_2, y_2), \qquad V_2 = (x_2, y_1), \qquad V_3 = (x_1, y_1)$$

denote the vertices of R^*. Furthermore let ℓ_{ij} denote the straight line through the vertices V_i and V_j where $i = 0, \ldots, 3$ and $j = i + 1 \bmod 4$. For a fixed point A_m let $Y_m \in R^*$ such that $d_m(R^*) = \|Y_m - A_m\|_1$ and define

$$\mathcal{J}_{ij} = \{ m \in \{1, \ldots, n\} : Y_m \in \ell_{ij} \}$$

where $i = 0, \ldots, 3$ and $j = i + 1 \bmod 4$.

If each line ℓ_{ij} contains at least one of the fixed points, then there is nothing to show. Therefore, assume that the straight line ℓ_{ij} does not contain any fixed point and define

$$l_{ij}(s) := \left\{ \begin{pmatrix} x+s \\ y+s \end{pmatrix} : \begin{pmatrix} x \\ y \end{pmatrix} \in \ell_{ij} \right\};$$

that is, $l_{ij}(s)$ is a parallel translation of the straight line l_{ij}. Since R^* is a minisum rectangle it is also an optimal solution to the line location problem

$$\min_{s \in \mathbb{R}} f_{ij}(s) = \sum_{m \in \mathcal{J}_{ij}} \omega_m \min\{\ell_1(Y - A_m) : Y \in \ell_{ij}(s)\}$$

$$= \begin{cases} \sum_{m \in \mathcal{J}_{ij}} \omega_m |s + y_2 - b_{m2}|; & i = 0, j = 1 \\ \sum_{m \in \mathcal{J}_{ij}} \omega_m |s + x_2 - a_{m2}|; & i = 1, j = 2 \\ \sum_{m \in \mathcal{J}_{ij}} \omega_m |s + y_1 - b_{m2}|; & i = 2, j = 3 \\ \sum_{m \in \mathcal{J}_{ij}} \omega_m |s + x_1 - a_{m2}|; & i = 3, j = 1 \end{cases}$$

Since a minimizer of $f_{ij}(s)$ has to be a weighted median, it follows that $s = 0$ cannot be a unique optimal solution and there has to exist $s \neq 0$ such that the straight line $\ell_{ij}(s)$ contains a fixed point. But then the rectangle defined by R^* and $\ell_{ij}(s)$ is also a minisum rectangle having at least one side on a direction line. In this way we can obtain a minisum rectangle having all four sides on direction lines. This implies the assertion.   □

There are $\mathcal{O}(M^2)$ intersection points of two distinct direction lines. For each intersection point V there are at most $(M-1)^2$ rectangles $R \in \mathscr{R}$ having their bottom left vertex at V and the remaining vertices at other intersection points. Hence, there exist $\mathcal{O}(M^4)$ rectangles satisfying the incidence property of Theorem 5.40. One of these rectangles is a minisum rectangle. Therefore, a finite dominating set is induced by Theorem 5.40. Using a brute-force algorithm, this FDS can be evaluated in $\mathcal{O}(M^5)$ operations.

The following example shows that Theorem 5.40 only states a sufficient but not a necessary condition; that is, an optimal rectangle does not need to have vertices at intersection points of direction lines.

Example 5.41. Consider the eight fixed points depicted in Fig. 5.7 where

$$A_1 = (-2.1, 0), \quad A_2 = (-1.9, 0), \quad A_3 = (0, 1.9), \quad A_4 = (0, 2.1),$$
$$A_5 = (0, -2.1), \quad A_6 = (0, -1.9), \quad A_7 = (1.9, 0), \quad A_8 = (2.1, 0).$$

For each fixed point assign a weight $\omega = 1$. Then each rectangle R with external fixed points A_1, A_4, A_5, A_8 and internal fixed points A_2, A_3, A_6, A_7 is an optimal solution with objective value $f(R) = 0.8$. In Fig. 5.7 the set of optimal rectangles is marked by a gray area; each rectangle which is contained in this area is an optimal solution.

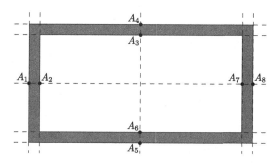

Fig. 5.7 Illustration of Example 5.41

5.6 Concluding Remarks

The results of this chapter show how to use the theory of previous chapters in order to obtain results for related problems. By transferring properties of the point–circle distance to the point–rectangle distance it is possible to obtain discretization results for the rectangle location problems considered in this chapter. These results provide a bridge between continuous optimization and discrete optimization. Using the geometric description of minisum rectangles it should be possible to develop more efficient algorithms for rectangle location problems. For the rectangle location problem with feasible set $\mathcal{R}_a$ a simple idea is already stated that reduces the complexity of Algorithm 5.1 from $\mathcal{O}(M^4 \log M)$ to $\mathcal{O}(M^3 \log M)$. Similar approaches seem to be possible for the problems with feasible sets $\mathcal{R}_p$ and $\mathcal{R}$.

The converse problem to the rectangle location problem studied in this chapter is also of interest. Which properties remain valid if weighted Manhattan norms are used in order to define diamonds and the Tschebyschow norm is used to measure distances? From a geometric point of view, also a generalization of the problem to the location of axis-parallel parallelotopes in any dimension $n \geq 2$ is a possible line of further research. For this problem it is also of interest to use a polyhedral norm having an axis-parallel parallelotope as unit ball for distance measuring.

Further interesting rectangle location problems include the problem of locating a rectangle with equal area. A first analysis of this problem has shown that this problem leads to a nonlinear global optimization problem. The theory presented in this chapter can be used in order to reduce the equal area problem to a finite number of convex optimization problems. The minimax version of the rectangle location problems with feasible sets $\mathcal{R}_a$ and $\mathcal{R}$ is studied by Abellanas et al. [AHI+03]. Further related problems are considered in [BBDG97, BG02, CS07, SV01].

Chapter 6
Extensions

In previous chapters hypersphere location problems have been studied where the objective consists of minimizing the weighted distances between a finite number of fixed points and a hypersphere. In Chapter 4 a generalization of the basic model has been considered where two unequal norms are used in the problem definition, and in Chapter 5 the special case of weighted Tschebyschow norms in the Manhattan plane has been discussed. A lot of extensions of these problems are possible. In the concluding remarks of previous chapters some possible extensions and lines of further research were already mentioned. In the following two extensions are discussed in more detail.

Positive and Negative Weights

Except for the weights ω_m associated to the fixed points in $\mathscr{D}$ we assume the same setting as in Chapter 3. For the weight ω_m of a fixed point A_m we allow a positive or a negative value. Obviously, the case $\omega_m = 0$ is not of interest. The problem considered in the following is called *minisum hypersphere problem with positive and negative weights*.

For the sake of distinction between fixed points with positive weights and fixed points with negative weights the following notation is introduced.

Notation 6.1. Given a set of fixed points $\mathscr{D} \subseteq \mathbb{R}^n$ let

$$\mathscr{D}^+ = \{A_m \in \mathscr{D} : \omega_m > 0\}, \qquad \mathscr{D}^- = \{A_m \in \mathscr{D} : \omega_m < 0\}.$$

It turns out that properties of the present problem rely on the weights associated to the fixed points. Therefore, we introduce the following notation.

M.-C. Körner, *Minisum Hyperspheres*, Springer Optimization and Its Applications 51, DOI 10.1007/978-1-4419-9807-1_6, © Springer Science+Business Media, LLC 2011

Notation 6.2. Let W denote the sum of positive and negative weights associated to the fixed points in $\mathscr{D}$, i.e.,

$$W := \sum_{m=1}^{M} \omega_m.$$

If $\mathscr{D}^- = \emptyset$ then the present problem is equivalent to the minisum hypersphere problem of Chapter 3. The case $\mathscr{D}^+ = \emptyset$, i.e., $\mathscr{D} = \mathscr{D}^-$, leads to a trivial, unbounded maximization problem. In the following it is assumed that both D^+ and D^- are not empty. Properties of the minisum hypersphere problem with positive and negative weights may differ strongly from the minisum hypersphere problem with positive weights. Due to the negative weights, the objective value of the present problem is not bounded below in general and instances exist where the objective value can get arbitrarily small. For the corresponding Weber problem with positive and negative weights (also known as semiobnoxious facility location problem) the following result is known which goes back to Drezner and Wesolowsky.

Lemma 6.3 ([DW91]). *For the Weber problem in $\mathbb{R}^n$ with positive and negative weights the following applies:*

- *If $W > 0$ then the optimal location is finite.*
- *If $W < 0$ then the optimal location is at infinity.*

Further results on the semiobnoxious Weber problem may be found for instance in [DW91, CHJT92, ND97, OT03]; surveys are given by Carrizosa and Plastria [CP99] as well as by Lozano and Mesa [LM00]. In the following it is shown that a similar result to Lemma 6.3 applies to the minisum hypersphere problem with positive and negative weights.

Theorem 6.4. *Let $\| \cdot \|$ be a norm in $\mathbb{R}^n$.*

- *If $W > 0$ then the objective function of the minisum hypersphere problem under norm $\| \cdot \|$ is bounded below*
- *if $W < 0$ then the objective function is not bounded below*

Proof. Let $W > 0$. We first show that any optimal hypersphere intersects the convex hull of the fixed points $\mathscr{D}$. To this end, let $S(X,r)$ be a hypersphere such that

$$\text{conv}(\mathscr{D}) \cap S(X,r) = \emptyset.$$

We distinguish the cases $\mathscr{D} \subseteq B(X,r)$ and $\mathscr{D} \cap B(X,r) = \emptyset$ where $B(X,r)$ denotes the closed ball with center X and radius r. In the former case there exists $r' < r$ such that $\mathscr{D} \subseteq B(X,r')$. Hence, we obtain

$$f(S(X,r)) - f(S(X,r')) = (r-r') \sum_{m=1}^{M} \omega_m = (r-r')W > 0.$$

In the latter case there exists $r' > r$ such that $\mathcal{D} \cap B(X,r') = \emptyset$. In a similar fashion it may be obtained $f(S(X,r)) - f(S(X,r')) > 0$. Hence, in both cases it can be concluded that any optimal solution has to intersect the convex hull of the fixed points $\mathcal{D}$. Thus for any $S(X,r) \in \mathcal{G}$ we have

$$f(S(X,r)) > \sum_{m:A_m \in \mathcal{D}^-} \omega_m d_m(S(X,r)) \geq \operatorname{diam}(\mathcal{D}) \sum_{m:A_m \in \mathcal{D}^-} \omega_m; \qquad (6.1)$$

that is, $f(S(X,r))$ is bounded below.

Now, let $W < 0$. We construct a series of hyperspheres whose objective values go to $-\infty$. To this end, let $W < 0$ and let $S(X,r)$ be a hypersphere such that

$$\operatorname{conv}(\mathcal{D}) \cap S(X,r) = \emptyset, \quad \mathcal{D} \subseteq B(X,r).$$

For all $r' > r$ we have

$$f(S(X,r')) - f(S(X,r)) = (r' - r) \sum_{m=1}^{M} \omega_m = (r' - r)W < 0.$$

Thus, we may improve the objective function by increasing r, i.e.,

$$\lim_{r \to \infty} f(S(X,r)) = -\infty.$$

$\square$

Remark 6.5. Based on Theorem 6.4 the minisum hypersphere problem with positive and negative weights can be studied for the case $W > 0$.

Extensive Facilities

Under the keyword *extensive facilities* a lot of problems have been studied in the literature where a new facility has to be located which cannot be represented by a point. Examples for this situation include the problem of locating straight lines (cf., e.g., [Sch99]), line segments (cf., e.g., [AEST93, ILY92, ES94]), and regular rectangular arrays of grid points (cf., e.g., [Pla91, RBFT99]). Also the minisum hypersphere problem may be considered as an extensive facility location problem. In the following we analyze the minisum hypersphere problem in the situation where also the fixed points are extensive, i.e., in the following we substitute the fixed points in $\mathcal{D}$ by closed and bounded subsets $\mathscr{A}_m \subseteq \mathbb{R}^n$. As in previous chapters we assume that $M \geq 2$ sets $\mathscr{A}_m$ are given. Furthermore, we assume that a positive weight $\omega_m > 0$ is associated to each set. We use the denotation *demand regions* in order to refer to the sets $\mathscr{A}_m$, $1 \leq m \leq M$. $\mathcal{D}$ is in the following the set which contains all demand regions. Similar to the point–hypersphere distance we define the distance between a hypersphere and a demand region.

Definition 6.6. Let $\| \cdot \|$ be a norm in $\mathbb{R}^n$, let $S = S(X,r)$ be a hypersphere, and $\mathscr{A}_m \in \mathscr{D}$ a demand region. The distance between S and $\mathscr{A}_m$ is given as

$$d(S, \mathscr{A}_m) = \min\{\|Z - Y\| : Y \in \mathscr{A}_m, Z \in S\}.$$

As in previous chapters, we use the abbreviation $d_m(S)$ in order to refer to $d(S, \mathscr{A}_m)$, $1 \le m \le M$. Using this notation the minisum hypersphere problem discussed in this part may be written as

$$\min f(S(X,r)) = \sum_{m=1}^{M} \omega_m d_m(S(X,r)).$$

Depending on the shape of the demand regions $\mathscr{A}_m$ it can be easy or very difficult to compute the distance between a hypersphere $S(X,r)$ and $\mathscr{A}_m$. Thus, the shape have a large impact on the solvability of the minisum hypersphere problem. A natural choice for the shape of the demand regions is a closed ball induced by the same norm $\| \cdot \|$ that we use to define the distance $d_m(R)$ and the set of hyperspheres $\mathscr{G}$.

For all $m = 1, \dots, M$, let

$$\mathscr{A}_m = B(A_m, r_m) := \{Y \in \mathbb{R}^n : \|A_m - Y\| \le r_m\}$$

where $A_m \in \mathbb{R}^n$ and $r_m > 0$; that is, each demand region $\mathscr{A}_m$ is a ball with center A_m and radius r_m. For such demand regions the following result applies.

Lemma 6.7. Let $\| \cdot \|$ be a norm in $\mathbb{R}^n$, let $S = S(X,r)$ be a hypersphere, and $\mathscr{A}_m \in \mathscr{D}$ a demand region such that $\mathscr{A}_m = B(A_m, r_m)$ for some $A_m \in \mathbb{R}^n$, $r_m > 0$. Then we have

$$d(S(X,r), \mathscr{A}_m) = \max\{0, d(S(X,r), A_m) - r_m\}.$$

Proof. Using Lemma 3.1 we obtain

$$d(S(X,r), A_m) - r_m = |\|X - A_m\| - r| - r_m.$$

We distinguish between four cases (see Fig. 6.1):

(i) $|\|X - A_m\| - r| < r_m$ and $\|X - A_m\| < r$.
(ii) $|\|X - A_m\| - r| < r_m$ and $\|X - A_m\| \ge r$.
(iii) $|\|X - A_m\| - r| \ge r_m$ and $\|X - A_m\| \ge r$.
(iv) $|\|X - A_m\| - r| \ge r_m$ and $\|X - A_m\| < r$.

We start with case (i) and define

$$Z = A_m + \frac{r - \|X - A_m\|}{\|X - A_m\|}(A_m - X).$$

Then

$$\|Z - X\| = \left\| \left(\frac{r - \|X - A_m\|}{\|X - A_m\|} + 1 \right)(A_m - X) \right\| = r,$$

i.e., $Z \in S(X,r)$. Furthermore, $\|Z - A_m\| = |r - \|X - A_m\|| < r_m$, i.e., $Z \in \mathscr{A}_m$. Hence, it follows $d(S(X,r), \mathscr{A}_m) = \max\{0, d(S(X,r), A_m) - r_m\} = 0$.

In case (ii) we define

$$Z = A_m + \frac{\|X - A_m\| - r}{\|X - A_m\|}(X - A_m)$$

and obtain analog to the previous case $d(S(X,r), \mathscr{A}_m) = 0$. Hence, $d(S(X,r), \mathscr{A}_m) = 0$, provided that

$$\|\|X - A_m\| - r| - r_m = d(S(X,r), A_m) - r_m < 0.$$

Now, we consider case (iii) and define

$$Y = A_m + \frac{r_m}{\|X - A_m\|}(X - A_m), \qquad Z = X + \frac{r}{\|X - A_m\|}(A_m - X).$$

Then $\|A_m - Y\| = r_m$ and $\|X - Z\| = r$; that is, $Y \in \mathscr{A}_m$ and $Z \in S(X,r)$. Hence,

$$d(S(X,r), \mathscr{A}_m) \leq \|Y - Z\|$$

$$= \left\|\left(\frac{r_m + r}{\|X - A_m\|} - 1\right)(X - A_m)\right\|$$

$$= |r_m + r - \|X - A_m\||$$

$$= |r_m - |\|X - A_m\| - r||$$

$$= |\|X - A_m\| - r| - r_m.$$

Choose any $Y' \in \mathscr{A}_m$, $Z' \in S(X,r)$. Then,

$$\|Y' - Z'\| \geq \|X - A_m\| - \|X - Z'\| - \|Y' - A_m\|$$

$$= \|X - A_m\| - r - r_m$$

$$= |\|X - A_m\| - r| - r_m.$$

Hence, in case (iii) we have $d(S(X,r), \mathscr{A}_m) = d(S(X,r), A_m) - r_m$. Defining

$$Y = A_m + \frac{r_m}{\|X - A_m\|}(A_m - X), \qquad Z = X + \frac{r}{\|X - A_m\|}(A_m - X),$$

we obtain the same result for case (iv). Summarizing, we obtain the assertion:

$$d(S(X,r), \mathscr{A}_m) = \max\{0, d(S(X,r), A_m) - r_m\}.$$

$\square$

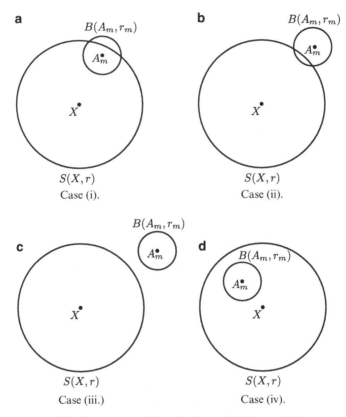

Fig. 6.1 Illustration of the four cases considered in the proof of Lemma 6.7

Remark 6.8. Note that the minisum hypersphere problem with demand regions is similar to the minisum hypersphere problem discussed in Chapter 3. In particular, the objective value of the latter problem is an upper bound on the objective value of the former problem. It seems possible to generalize many properties to the minisum hypersphere problem with demand regions. For instance, it is clear that any optimal solution has to touch at least one demand region.

References

[AEST93] P.K. Agarwal, A. Efrat, M. Sharir, and S. Toledo. Computing a segment center for a planar point set. *Journal of Algorithms*, 15:314–323, 1993.

[AHI+03] M. Abellanas, F. Hurtado, C. Icking, L. Ma, B. Palop, and P.A. Ramos. Best fitting rectangles. Technical report, Fernuniversität Hagen, 2003.

[All78] P.D. Allison. Measures of inequality. *American Sociological Review*, 43:865–880, 1978.

[AT05] L. An and P. Tao. The dc (difference of convex functions) programming and dca revisited with dc models of real world nonconvex optimization problems. *Annals of Operations Research*, 133:23–46, 2005.

[AW04] A. Atieg and G.A. Watson. Fitting circular arcs by orthogonal distance regression. *Applied Numerical Analysis & Computational Mathematics*, 1:66–76, 2004.

[Baj88] C. Bajaj. The algebraic degree of geometric optimization problems. *Discrete and Computational Geometry*, 3:117–191, 1988.

[BBDG97] G. Barequet, A.J. Briggs, M.T. Dickerson, and M.T. Goodrich. *Algorithms and Data Structures*, chapter Offset-Polygon Annulus Placement Problems. Springer, 1997.

[BC86] M. Berman and D. Culpin. The statistical behaviour of some least squares estimators of the centre and radius of a circle. *Journal of the Royal Statistical Society. Series B (Methodological)*, 48:183–196, 1986.

[BCH09] R. Blanquero, E. Carrizosa, and P. Hansen. Locating objects in the plane using global optimization techniques. *Mathematics of Operations Research*, 34:837–858, 2009.

[Ber87] M. Berger. *Geometry II*. Springer, corrected edition, 1987.

[BG02] M. Bern and D. Goldberg. Paper position sensing. In *SCG '02: Proceedings of the eighteenth annual symposium on Computational geometry*, pages 74–81, 2002.

[BJKS09] J. Brimberg, H. Juel, M. Körner, and A. Schöbel. Locating a general minisum circle on the plane. Technical Report 10, Institute for Numerical and Applied Mathematics, University of Göttingen, 2009.

[BJS02] J. Brimberg, H. Juel, and A. Schöbel. Linear facility location in three dimensions: Models and solution methods. *Operations Research*, 50:1050–1057, 2002.

[BJS07] J. Brimberg, H. Juel, and A. Schöbel. Locating a circle on a sphere. *Operations Research*, 55:782–791, 2007.

[BJS09a] J. Brimberg, H. Juel, and A. Schöbel. Locating a circle on the plane using the minimax criterion. *Studies in Location Analysis*, 17:45–60, 2009.

[BJS09b] J. Brimberg, H. Juel, and A. Schöbel. Locating a minisum circle in the plane. *Discrete Applied Mathematics*, 157:901–912, 2009.

[BK90] O. Berman and E.H. Kaplan. Equity maximizing facility location schemes. *Transportation Science*, 24:137–144, 1990.

[BKOS00] M. de Berg, M. van Kreveld, M. Overmars, and O. Schwarzkopf. *Computational Geometry: Algorithms and Applications*. Springer, second edition, 2000.

M.-C. Körner, *Minisum Hyperspheres*, Springer Optimization and Its Applications 51, 109
DOI 10.1007/978-1-4419-9807-1, © Springer Science+Business Media, LLC 2011

[BLR76]   E.D. Brill, J.C. Liebman, and C.S. ReVelle. Equity measures for exploring water qual-
           ity management alternatives. *Water Resources Research*, 12:845–851, 1976.

[BSM98]   V. Boltyanski, V. Soltan, and H. Martini. *Geometric Methods and Optimization Prob-
           lems (Combinatorial Optimization)*. Springer, 1998.

[Cha65]   N.N. Chan. On circular functional relationships. *Journal of the Royal Statistical So-
           ciety. Series B (Methodological)*, 27:45–56, 1965.

[Cha00]   T.M. Chan. Approximating the diameter, width, smallest enclosing cylinder, and
           minimum-width annulus. In *SCG '00: Proceedings of the sixteenth annual sympo-
           sium on Computational geometry*, pages 300–309, 2000.

[CHJT92]   P.C. Chen, P. Hansen, B. Jaumard, and H. Tuy. Weber's problem with attraction and
           repulsion. *Journal of Regional Science*, 32:467–486, 1992.

[CL05]   N. Chernov and C. Lesort. Least squares fitting of circles. *Journal of Mathematical
           Imaging and Vision*, 23:239–252, 2005.

[CLT05]   Y.T. Chan, B.H. Lee, and S.M. Thomas. Approximate maximum likelihood estimation
           of circle parameters. *Journal of Optimization Theory and Applications*, 125:723–734,
           2005.

[CMR96]   A. Corbalan, M.L. Mazón, and T. Recio. Geometry of bisectors for strictly convex dis-
           tances. *International Journal of Computational Geometry and Applications*, 6:45–58,
           1996.

[Coo19]   J.L. Coolidge. The circle nearest to n given points, and the point nearest to n given
           circles. *The Annals of Mathematics, Second Series*, 21:94–97, 1919.

[CP99]   E. Carrizosa and F. Plastria. Location of semi-obnoxious facilities. *Studies in Loca-
           tional Analysis*, 12:1–27, 1999.

[CS07]   D. Chaudhuri and A. Samal. A simple method for fitting of bounding rectangle to
           closed regions. *Pattern Recognition*, 40:1981–1989, 2007.

[CS08]   N. Chernov and P.N. Sapirstein. Fitting circles to data with correlated noise. *Compu-
           tational Statistics & Data Analysis*, 52:5328–5337, 2008.

[Day47]   M.M. Day. Some characterizations of inner-product spaces. *Transactions of the Amer-
           ican Mathematical Society*, 62:320–337, 1947.

[DBMS02]   J. M. Diaz-Bánez, J. A. Mesa, and A. Schöbel. Continuous location of dimensional
           structures. *European Journal of Operational Research*, 152:22–44, 2002.

[DDG09]   T. Drezner, Z. Drezner, and J. Guyse. Equitable service by a facility: Minimizing the
           gini coefficient. *Computers & Operations Research*, 36:3240–3246, 2009.

[Dic85]   D. R. Dicks. *Early Greek Astronomy to Aristotle (Aspects of Greek and Roman Life
           Series)*. Cornell University Press, 1985.

[DKSW02]   Z. Drezner, K. Klamroth, A. Schöbel, and G.O. Wesolowsky. The Weber problem. In
           Z. Drezner and H.W. Hamacher, editors, *Facility Location: Applications and Theory*,
           pages 1–36. Springer, 2002.

[DLM07]   L.D. Drager, J.M. Lee, and C.F. Martin. On the geometry of the smallest circle en-
           closing a finite set of points. *Journal of the Franklin Institute*, 344:929–940, 2007.

[DM85]   R. Durier and C. Michelot. Geometrical properties of the Fermat-Weber problem.
           European Journal of Operational Research, 20:332–343, 1985.

[DSW02]   Z. Drezner, G. Steiner, and G.O. Wesolowsky. On the circle closest to a set of points.
           Computers and Operations Research, 29:637–650, 2002.

[DW91]   Z. Drezner and G.O. Wesolowsky. The Weber problem on the plane with some nega-
           tive weights. *Information Systems and Operational Research*, 29:87–99, 1991.

[ES94]   A. Efrat and M. Sharir. A near-linear algorithm for the planar segment center problem.
           In *SODA '94: Proceedings of the fifth annual ACM-SIAM symposium on Discrete
           algorithms*, pages 87–97. Society for Industrial and Applied Mathematics, 1994.

[FC94]   F.T Farago and M.A. Curtis. *Handbook of dimensional measurement*. Industrial Press
           Inc., 3rd edition, 1994.

[FS04]   S. Fujita and T. Suzuki. A study on the optimal pattern of radial-ring high-speed
           network and modal split. *Journal of the City Planning Institute of Japan*, 39:835–840,
           2004.

[GG99]   T.W. Gamelin and R.E. Greene. *Introduction to Topology*. Dover Publications, second edition, 1999.

[GHT09]   O. Gluchshenko, H.W. Hamacher, and A. Tamir. An optimal O(nlogn) algorithm for finding an enclosing planar rectilinear annulus of minimum width. *Operations Research Letters*, 37:168–170, 2009.

[GJ79]   M.R. Garey and D.S. Johnson. *Computers and Intractability, A Guide to the Theory of NP-Completeness*. W.H. Freeman and Company, 1979.

[GLRS98]   J. García-López, P.A. Ramos, and J. Snoeyink. Fitting a set of points by a circle. *Discrete and Computational Geometry*, 20:389–402, 1998.

[Glu08]   O. Gluchshenko. *Annulus and Center Location Problems*. PhD thesis, TU Kaiserslautern, 2008. http://kluedo.ub.uni-kl.de/frontdoor.php?source_opus=2276.

[Gri91]   W. Grimson. *Object Recognition by Computer*. The MIT Press, 1991.

[Gro69]   M.L. Gromov. On simplexes inscribed in a hypersurface. *Mathematical Notes*, 5:52–56, 1969. (Translation from Matematicheskie Zametki, Vol. 5, pp. 81–89, 1969).

[Haz90]   M. Hazewind, editor. *Geometry of Banach spaces, duality mappings and nonlinear problems*. Kluwer, 1990.

[HD04]   H.W. Hamacher and Z. Drezner. *Facility Location. Application and Theory*. 2nd edition. Springer, Berlin, 2004.

[Hob13]   E. W. Hobson. *Squaring the circle - a history of the problem*. Cambridge University Press, 1913.

[Hor00]   Á.G. Horváth. On bisectors in Minkowski normed spaces. *Acta Mathematica Hungarica*, 89:233–246, 2000.

[IKLM94]   C. Icking, R. Klein, N.-M. Lê, and L. Ma. Convex distance functions in 3-space are different. *Fundamenta Informaticae*, 22:116–123, 1994.

[IKM+99]   C. Icking, R. Klein, L. Ma, S. Nickel, and A. Weißler. On bisectors for different distance functions. In *SCG '99: Proceedings of the fifteenth annual symposium on Computational Geometry*, pages 291–299, 1999.

[ILY92]   H. Imai, D.T. Lee, and C.D. Yang. 1-segment center problems. *INFORMS Journal on Computing*, 4:426–434, 1992.

[KBJS09]   M. Körner, J. Brimberg, H. Juel, and A. Schöbel. General minisum circle location. In *Proceedings of the 21st Canadian Conference on Computational Geometry*, pages 111–114, 2009.

[KBJS10]   M. Körner, J. Brimberg, H. Juel, and A. Schöbel. Geometric fit of a point set by generalized circles. *J. Global. Optim.*, 2010.

[KM90]   N. Korneenko and H. Martini. Approximating finite weighted point sets by hyperplanes. In *SWAT '90: Proceedings of the 2nd Scandinavian Workshop on Algorithm Theory*, pages 276–286, 1990.

[KM93]   N. Korneenko and H. Martini. Hyperplane approximation and related topics. In J. Pach, editor, *New Trends in Discrete and Computational Geometr*, pages 135–162. Springer, 1993.

[KN73]   H. Kramer and A.B. Németh. Equally spaced points for families of compact sets in Minkowski spaces. *Mathematica (Cluj)*, 15:71–78, 1973.

[Kno93]   W. R. Knorr. *The ancient tradition of geometric problems*. Dover Publications, 1993.

[KS10]   M. Körner and A. Schöbel. Weber problems with high-speed lines. *TOP*, 18:223–241, 2010.

[LL91]   V.B. Le and D.T. Lee. Out-of-roundness problem revisited. *IEEE Transactions on Pattern Analysis and Machine Intelligence*, 13:217–223, 1991.

[LM00]   A.J. Lozano and J.A. Mesa. Location of facilities with undesirable effects and inverse location problems: a classification. *Studies in Locational Analysis*, 14:253–291, 2000.

[Lue04]   D.G. Luenberger. *Linear and nonlinear programming*. Kluwer, second edition, 2004.

[Ma00]   L. Ma. *Bisectors and Voronoi Diagrams for Convex Distance Functions*. PhD thesis, Fernuniversität Hagen, 2000. http://wwwpi6.fernuni-hagen.de/Publikationen/tr267.

[Meg82]   N. Megiddo. Linear-time algorithms for linear programming in R3 and related problems. In *SFCS '82: Proceedings of the 23rd Annual Symposium on Foundations of Computer Science*, pages 329–338. IEEE Computer Society, 1982.

[Min11]   H. Minkowski. *Theorie der konvexen Körper*, volume 2 of *Gesammelte Abhandlungen*. Teubner, Berlin, 1911. In German.

[MS94]   M.T. Marsh and D.A. Schilling. Equity measurement in facility location analysis: A review and framework. *European Journal of Operational Research*, 74:1–17, 1994.

[MS98]   H. Martini and A. Schöbel. Median hyperplanes in normed spaces - a survey. *Discrete Applied Mathematics*, 89:181–193, 1998.

[MS99]   H. Martini and A. Schöbel. Two characterizations of smooth norms. *Geometriae Dedicata*, 77:173–183, 1999.

[MS01]   H. Martini and A. Schöbel. Median and center hyperplanes in Minkowski spaces - a unified approach. *Discrete Mathematics*, 241:407–426, 2001.

[MS04]   H. Martini and K. Swanepoel. The geometry of Minkowski spaces - a survey. Part II. *Expositiones Mathematicae*, 22:93–144, 2004.

[MSW01]   H. Martini, K. Swanepoel, and G. Weiss. The geometry of Minkowski spaces - a survey. Part I. *Expositiones Mathematicae*, 19:97–142, 2001.

[MSW02]   H. Martini, K. Swanepoel, and G. Weiss. The fermat-torricelli problem in normed planes and spaces. *Journal of Optimization Theory and Applications*, 115:238–314, 2002.

[Mul91]   G.F. Mulligan. Equality measures and facility location. *Papers in Regional Science*, 70:345–365, 1991.

[ND97]   S. Nickel and E.M. Dudenhöffer. Weber's problem with attraction and repulsion under polyhedral gauges. *Journal of Global Optimization*, 11:409–432, 1997.

[Nic95]   S. Nickel. *Discretization of Planar Location Problems*. Shaker, 1995.

[Nie02]   Y. Nievergelt. A finite algorithm to fit geometrically all midrange lines, circles, planes, spheres, hyperplanes, and hyperspheres. *Numerische Mathematik*, 91:257–303, 2002.

[Nie04]   Y. Nievergelt. Perturbation analysis for circles, spheres, and generalized hyperspheres fitted to data by geometric total least-squares. *Mathematics of Computation*, 73:169–180, 2004.

[Nie10]   Y. Nievergelt. Median spheres: theory, algorithms, applications. *Numerische Mathematik*, 114:573–606, 2010.

[NP05]   S. Nickel and J. Puerto. *Location Theory, A Unified Approach*. Springer, 2005.

[Ogr09]   W. Ogryczak. Equity measures for exploring water quality management alternatives. *Annals of Operations Research*, 167:61–86, 2009.

[OT03]   Y. Ohsawa and K. Tamura. Efficient location for a semi-obnoxious facility. *Annals of Operations Research*, 123:173–188, 2003.

[PC01]   F. Plastria and E. Carrizosa. Gauge distances and median hyperplanes. *Journal of Optimization Theory and Applications*, 110:173–182, 2001.

[Pea74]   C.E.M. Pearce. Locating concentric ring roads in a city. *Transportation Science*, 8:142–168, 1974.

[Phe89]   R.R. Phelps. *Convex Functions, Monotone Operators and Differentiability*. Number 1364 in Lecture Notes in Mathematics. Springer, 1989.

[Pla91]   F. Plastria. Optimal gridpositioning or single facility location on the torus. *Recherche opérationnelle*, 25:19–29, 1991.

[Pla02]   F. Plastria. Continuous covering location problems. In Z. Drezner and H.W. Hamacher, editors, *Facility Location: Applications and Theory*, pages 37–79. Springer, 2002.

[RBFT99]   M. Rayco, M. Brenda, R.L. Francis, and A. Tamir. A p-center grid-positioning aggregation procedure. *Computers & Operations Research*, 26:1113–1124, 1999.

[Roc70]   R.T. Rockafellar. *Convex Analysis*. Princeton University Press, 1970.

[RS87]   G. Robins and C. Shute. *The Rhind mathematical papyrus. An ancient Egyptian text*. British Museum, 1987.

[Sch99]   A. Schöbel. *Locating Lines and Hyperplanes, Theory and Algorithms*. Kluwer Academic Publishers, Dordrecht, 1999.

[Sch07]   D. Scholz. Theorie und geometrische Lösungsverfahren zur Platzierung von Kreisen und Kugeln. Master's thesis, University Göttingen, 2007. In German.

[SKW02]   B. Schandl, K. Klamroth, and M.M. Wiecek. Norm-based approximation in multicriteria programming. *Computers & Mathematics with Applications*, 44:925–942, 2002.

[SLW95]   K. Swanson, D. T. Lee, and V.L. Wu. An optimal algorithm for roundness determination on convex polygons. *Computational Geometry: Theory and Applications*, 5:225–235, 1995.

[SS10]   A. Schöbel and D. Scholz. The big cube small cube solution method for multidimensional facility location problems. *Computers & Operations Research*, 37:115–122, 2010.

[Sun09]   T.H. Sun. Applying particle swarm optimization algorithm to roundness measurement. *Expert Systems with Applications*, 36:3428–3438, 2009.

[SV01]   H. Suesse and K. Voss. A new efficient algorithm for fitting rectangles and squares. In *Proc. IEEE International Conference on Image Processing*, pages 809–812, 2001.

[Thi87]   J.-F. Thisse. Location theory, regional science, and economics. *Journal of Regional Science*, 27:519–528, 1987.

[Tho96]   A.C. Thompson, editor. *Minkowski Geometry*. Cambridge University Press, 1996.

[WCM93]   A. Wallack, J. Canny, and D. Manocha. Object localization using crossbeam sensing. In *IEEE International Conference on Robotics and Automation*, 1993.

[Web09]   A. Weber. Über den Standort der Industrien. *Tübingen*, 1909. In German.

[Wit64]   C. Witzgall. Optimal location of a central facility: mathematical models and concepts. Technical Report 8388, National Bureau of Standards, 1964.

[WWR85]   J.E. Ward, R.E. Wendell, and E. Richard. Using block norms for location modeling. *Operations Research*, 33:1074–1090, 1985.

Index